Lecture Notes in Mathematics

Edited by A. Dold and B. Eckmann

1386

A. Bellen C.W. Gear E. Russo (Eds.)

Numerical Methods for Ordinary Differential Equations

Proceedings of the Workshop
held in L'Aquila (Italy), Sept. 16–18, 1987

Springer-Verlag

Berlin Heidelberg New York London Paris Tokyo Hong Kong

Editors

Alfredo Bellen
Dipartimento di Scienze Matematiche, Università di Trieste
34100 Trieste, Italy

Charles W. Gear
Department of Computer Science, University of Illinois
Urbana, IL 61801, USA

Elvira Russo
Dipartimento di Informatica e Applicazioni, Università di Salerno
84100 Salerno, Italy

Mathematics Subject Classification (1980): 65L05, 65J05, 65Q05, 65W05, 65M99

ISBN 3-540-51478-3 Springer-Verlag Berlin Heidelberg New York
ISBN 0-387-51478-3 Springer-Verlag New York Berlin Heidelberg

Printing and binding: Druckhaus Beltz, Hemsbach/Bergstr.
2146/3140-543210 – Printed on acid-free paper

INTRODUCTION.

The theory of the numerical integration of initial value problems is by now classical. This theory has formed the basis for a spectrum of good codes for stiff and non stiff problems that enable users of conventional serial computers to solve the majority of problems with no great difficulty or inefficiency. Consequently, the thrust of much current research has been to extend the classical results to more general cases, especially to enable the application of ODE methods to problems such as those arising from PDEs by the application of the method of lines, a method that is particularly easy to apply in an automatic code, and to the differential-algebraic, integral, and delay equations that frequently arise in computer modeling of physical systems. As the number of users increases as well as their need for the accurate and rapid solution of the larger and larger systems that can be generated automatically with today's computer systems, two other important thrusts emerge. One is the improvement of codes to meet the needs of professionals in other disciplines who do not wish to be concerned with technical details of numerical methods. The second is the development of ODE methods suitable for the emerging generations of parallel computers.

The purpose of the symposium on Ordinary Differential Equations organized by A.Bellen, A.Pasquali, L.Pasquini, E.Russo and D.Trigiante at the University of L'Aquila, Italy on September 16-18, 1987 was to examine some of these new developments and to explore the connections betweeen the classical background and new research areas. Eight invited speakers gave fourteen one-hour talks. Material from these talks is presented in the following chapters.

The organizers of the symposium and the editors of this volume wish to express their thanks and appreciation for the support provided for this meeting by the "Dipartimento di Matematica Pura ed Applicata" and the "Dipartimento di Energetica" of the University of L'Aquila and by the Italian National Council of Research (CNR).

This set of papers is by no means intended to summarize the vast range of work that is occuring today on important new problems related to ODEs, but to highlight some problem areas. There are still many challenging areas of investigation to pursue. In the paragraphs below we mention a few that the editors find particularly interesting and believe to be important. The papers in this volume address some of them.

Although codes are well developed, we can still expect to see improvements in a number of areas. The user would like a code in which the tolerance control parameter is proportional to the global error. Even more, the user would like an estimate of the global error after execution. Efficient techniques for this require further development. In problems with frequent discontinuities most codes are somewhat inefficient (codes not designed to expect such occurences may be extremely inefficient). The effective detection of discontinuities needs further development, but there has been progress on the related problem of the cost of restarting a multistep method after a discontinuity. At the moment this is done as for an initial start. It is interesting to wonder if there are techniques that could treat a discontinuity as an impulse and project the history of the solution through the discontinuity, avoiding a return to low order.

Most codes compute discrete values or at most piecewise polynomial approximations that are no better than continuous. Recent code developments have introduced efficient interpolants with superior smoothness and accuracy characteristics.

Extensions of ODE methods to classes of functional differential equations is also an active area. Smooth interpolants are important in these problems. In integral and delay equations some codes have been developed and there has been work on the stability of methods. Tha main challenge is to find a good definition of stability and appropriate choices of test equations.

The embedding of ODE codes in packages for PDEs has made it possible to solve limited classes of time-dependent PDEs with minimum user interaction. However, the efficiency of these methods is not usually great so that they are not useful for large-scale problems. One area in which development is needed is the closer interface between automatic ODE solvers so that more complex boundaries can be accomodated and so that, for example, the spatial discretization can be changed during the integration.

Differential-algebraic equations have received extensive attention in the last decade and the theory is well developed. But, we still do not have methods for high-index problems, there are no general codes for problems of index greater than one, and the determination of initial values remains a difficulty.

The recent easy availability of parallel computers has excited the interest of researchers in ODEs, and a number of methods have been developed for small-scale parallelism. The problem of developing a code for a class of parallel computers as opposed to one for a particular architecture is far more difficult than that of developing a code for a serial computer, and such code development will undoubtedly lead to new advances in our understanding of parallel methods. The application of large-scale parallelism to ODEs efficiently is a challenge which has only recently begun to receive attention. Here the problem seems to be to develop completely new methods which overcome the inherent serialism of information flow forward in time in an initial value problem, and the communication costs of information flow betweeen different components in a large, stiff system. We espect this area to be a major research area in the next decade.

March, 1988

A.Bellen
C.Gear
E.Russo

TABLE OF CONTENTS

STABILITY IN LINEAR ABSTRACT DIFFERENTIAL EQUATIONS

by C. BAIOCCHI (*)
Dipartimento di Matematica
Università di Pavia
27100 Pavia - Italy

1. INTRODUCTION.

In the usual framework of the Hilbert triple $\{V, H, V^*\}$, we will deal with an abstract differential equation of the type:

$$u'(t) + A\,u(t) = f(t) \qquad \text{in } V^*, \quad \text{for } t > 0 \tag{1.1}$$

where A is a linear continuous operator from V to V*, which is coercive (more details later).

For (1.1) we want to study stability and convergence properties of linear multistep one leg methods; in other words we are given two polynomials ρ, σ of degree $r \geq 1$:

$$\rho(z) = \sum_{j=0}^{r} \alpha_j z^j \; ; \quad \sigma(z) = \sum_{j=0}^{r} \beta_j z^j \qquad (z \in \mathbb{C}) \tag{1.2}$$

and we replace (1.1) by the sequence of problems (for $n \geq 0$; $k > 0$ is the stepsize (1)):

$$\frac{1}{k} \sum_{j=0}^{r} \alpha_j u_{n+j} + A \sum_{j=0}^{r} \beta_j u_{n+j} = \sum_{j=0}^{r} \beta_j f_{n+j} \tag{1.3}$$

(*) This work was partially supported by 40% and 60% funds of the M.P.I., and partially supported by C.N.R., I.A.N. of Pavia.

(1) The extension of the results to the case of variable step size is an interesting open problem.

As a stability assumption on ρ, σ, we will impose the following property which, at a first glance, looks very restrictive:

There exists a polynomial q, *with* $\deg(q) \leq r$, *and there exist* r *polynomials,* $p_1, p_2, \ldots, p_r$, *with* $\deg(p_m) \leq r - 1$ $(m = 1, \ldots, r)$ *such that for all* $z, w \in \mathbb{C}$ *the following identity holds:*

(1.4) $$\rho(z)\,\overline{\sigma(w)} + \overline{\rho(w)}\,\sigma(z) + (1 - z\,\overline{w}) \sum_{m=1}^{r} p_m(z)\,\overline{p_m(w)} = q(z)\,\overline{q(w)}$$

Sometimes, for more precise results, we will also ask:

(1.5) *The* r *polynomials* $\{p_m\}$ *are linearly independent.*

Let us remark that from (1.4) with $w = z$ and $|z| \geq 1$, we obviously get:

(1.6) $$z \in \mathbb{C}, |z| \geq 1 \quad \Rightarrow \quad \operatorname{Re}\ \rho(z)\,\overline{\sigma(z)} \geq 0$$

so that, if we add the requirement:

(1.7) ρ *and* σ *are relatively prime*

we are in fact dealing with an A-stable couple (ρ, σ) (Dahlquist [3]). Conversely we will show (see Baiocchi-Crouzeix[2]):

(1.8) *Property* (1.6) *implies* (1.4); *furthermore, if* (1.7) *is satisfied, then also* (1.5) *holds true*

so that our stability results holds true for *any* A-stable method.

On the other hand, from (1.8) we will also get a new proof of the well known result:

(1.9) *A-stability is equivalent to G-stability*

(see Dahlquist [5]).

2. PROOF OF (1.8), (1.9).

In this section we assume that ρ and σ satisfy (1.6); we do not assume the coefficient α_j, β_j in (1.2) are real; r will be the true degree of ρ, σ (say $\alpha_r \neq 0$, so that, from (1.6) with large $|z|$, $\beta_r \neq 0$).

For the polynomial q(z) given by

$$(2.1) \qquad q(z) = \sum_{j=0}^{s} q_j z^j \qquad (z \in \mathbb{C})$$

we will denote by $\bar{q}$ the polynomial defined by $\bar{q}(z) = \overline{q(\bar{z})}$ (i.e. the coefficients of $\bar{q}$ are the conjugate of the corresponding coefficients of q). We start with the following Lemma:

LEMMA 2.1. *Let* E(z) *be defined by:*

$$(2.2) \qquad E(z) = \rho(z)\, \bar{\sigma}\left(\frac{1}{z}\right) + \bar{\rho}\left(\frac{1}{z}\right) \sigma(z) \qquad (z \in \mathbb{C}, z \neq 0)$$

There exists a polynomial q, *with* $\deg(q) \leq r$, *such that, for all* $z \in \mathbb{C}$ *with* $z \neq 0$:

$$(2.3) \qquad E(z) = q(z)\, \bar{q}\left(\frac{1}{z}\right)$$

Proof. Let us assume $E \not\equiv 0$ (otherwise we can choose $q \equiv 0$). From (2.2) easily follows:

$$(2.4) \qquad E(1/\bar{z}) = \overline{E(z)} \qquad (z \in \mathbb{C}, z \neq 0)$$

which in particular implies that, for $|z| = 1$, the values of E(z) are real; more precisely, from (1.6) we get:

$$(2.5) \qquad E(z) \geq 0 \qquad \text{for } |z| = 1$$

and from (2.2) follows:

(2.6) *for some* $n \in \mathbb{Z}$, $z^n E(z)$ *is a polynomial*

(e.g. (2.6) is true for $n \geq r$); in particular, from (2.6), we can work with the (non zero) "roots of E".

Let z_0 be a non zero root of E (if any!); from (2.4) we see that $(1/\overline{z_0})$ is also a root of E. We want to show that E can be factorized in the form:

$$(2.7) \qquad E(z) = \tilde{E}(z) \;.\; (z - z_0)(1/\bar{z} - \overline{z_0})$$

In fact (2.7) is obvious, because of (2.6), if $z_0 \neq 1/\overline{z_0}$. If, $z_0 = 1/\overline{z_0}$ (i.e. $|z_0| = 1$) we need to show that z_0 is at least a double root of E: this follows obviously from (2.5).

The function $\tilde{E}(z)$ defined by (2.7) is such that (2.6) still holds for $\tilde{E}$; and an easy computation shows that also (2.4), (2.5) are satisfied by $\tilde{E}$; so that if $\tilde{E}$ has non zero roots we can factorize $\tilde{E}$ in a form similar to (2.7). After a finite number of steps we must end up with (2) :

$$(2.8) \qquad E(z) = \tilde{\tilde{E}}(z) \cdot \Big[\prod_{j=1}^{s} (z - z_j)\Big] \cdot \Big[\prod_{j=1}^{s} (\tfrac{1}{z} - \bar{z}_j)\Big]$$

where $\tilde{\tilde{E}}(z)$ has the form $\lambda \,.\, z^m$, for suitable $\lambda \in \mathbb{C}$, $m \in \mathbb{Z}$; but $\tilde{\tilde{E}}(z)$ must still satisfy (2.4), which implies $m = 0$, say $\tilde{\tilde{E}}(z) = \lambda$; and must also satisfy (2.5), which implies that λ is real and non negative.

Setting

$$(2.9) \qquad q(z) = \sqrt{\lambda} \prod_{j=1}^{s} (z - z_j)$$

formula (2.8) becomes (2.3). Let us assume, by contradiction, that $s = \deg(q)$ is greater than r; multiplying (2.3) by z^s we get an identity between polynomials; the degree of left hand member is at most $s + r < 2\,s$; the degree of right hand member being 2 s we get a contradiction. ■

REMARK 2.1. If the coefficients α_j, β_j are real, for any z_1 complex root of E, we will also have $E(\overline{z_0}) = 0$ (and, by (2.5), $E(\frac{1}{z_0}) = 0$). Instead of (2.8) we will then write

$$E(z) = \tilde{E}(z)\, [(z - z_0)(z - \overline{z_0})] \,.\, [(\tfrac{1}{z} - \overline{z_0})(\tfrac{1}{z} - z_0)]$$

and we never loose reality of coefficients in the single step; so that we end up with a polynomial q with real coefficients. However, for α_j, β_j real, a shorter proof for Lemma 2.1 can be given; see [2].

---

(2) Let n_0 be the minimum value of n for which $z^n E(z)$ is a polynomial; and let N be the degree of $z^{n_0}E(z)$. It is easily seen that at each step $E \to \tilde{E}$, the value of N decreases exactly by 2. This argument is also an alternative proof of the fact that deg(p) will be at most r.

REMARK 2.2. The proof of Lemma 2.1 is of constructive nature: if we know the roots of E, we can exhibit an explicit formula for q (see (2.9)) by just splitting such roots between q(z) and $\bar{q}(1/z)$.

COROLLARY 2.1. *There exists an hermitian matrix* G:

$$(2.10) \qquad G \equiv (g_{j,i})_{j,i=1,\dots,r} \quad ; \qquad g_{j,i} = \overline{g_{i,j}}$$

such that, for all $z, w \in \mathbb{C}$:

$$(2.11) \qquad q(z)\,\bar{q}(w) - \rho(z)\,\bar{\sigma}(w) - \bar{\rho}(w)\,\sigma(z) = (1 - z\,w) \sum_{j;i=1}^{r} g_{j,i}\; z^{j-1}\, w^{i-1}$$

Proof. The left hand side of (2.11) is a polynomial in z,w, of degree at most r in z, w; and it vanishes identically for w =1/z because of (2.3); so that formula (2.11) must hold true for suitably choosen $(g_{j,i})_{j,i=1,\dots,r}$. On the other hand the polynomial $g(z, w) = \dfrac{q(z)\bar{q}(w) - \rho(z)\,\bar{\sigma}(w) - \bar{\rho}(w)\,\sigma(z)}{1 - z\,w}$ satisfies the identity $\overline{g(z, w)} = g(\bar{w}, \bar{z})$, so that $\overline{g_{j,i}} = g_{i,j}$, or $G = G^*$. ■

LEMMA 2.2. *Under assumption* (1.7) *the matrix* G *defined in* Cor. 2.1 *verifies:*

(2.12) G *is strictly positive definite*

Proof. Let us first remark that the subset of $\mathbb{C}$ given by $\{z \in \mathbb{C} \mid$ for some $\lambda \in \mathbb{C}$ z is a multiple root of $\rho + \lambda\,\sigma\}$ is finite: in fact from $\rho(z) + \lambda\,\sigma(z) = 0$, $\rho'(z) + \lambda\,\sigma'(z) = 0$ we get $\rho(z)\,\sigma'(z) - \rho'(z)\,\sigma(z) = 0$ so that our set has at most 2r - 1 elements. Now, from (1.7), we see that for at most a finite number of values of $\lambda \in \mathbb{C}$, the polynomial $\rho + \lambda\,\sigma$ can have multiple roots.

Fix any λ with

(2.13) $\lambda > 0$; $\rho + \lambda\,\sigma$ *has no multiple roots*

and denote by $z_1, z_2, \dots, z_r$ the r roots of $\rho + \lambda\,\sigma$, and by V the corresponding Vandermonde matrix; we will show:

(2.14) *The matrix* $M = V^*\,G\,V$ *is strictly positive definite*

and (2.12) will follow. Remark that, from $\rho(z_i) + \lambda\,\sigma(z_i) = 0$ $(i = 1, \dots, r)$ it follows:

(2.15) $\quad \rho(z_i)\,\overline{\sigma(z_j)} = -\lambda\,\sigma(z_i)\,\overline{\sigma(z_j)} \qquad (i, j = 1, \ldots, r)$

and in particular, from (2.11), the matrix M defined in (2.14) is given by:

(2.16) $\quad M \equiv (m_{j,i}); \quad m_{j,i} = \dfrac{1}{1 - z_j\,\overline{z_i}}\,[q(z_j)\,\overline{q(z_i)} + 2\,\lambda\,\sigma(z_j)\,\overline{\sigma(z_i)}]$

On the other hand, from $\rho(z_i) + \lambda\,\sigma(z_i) = 0$ and (1.7), it follows

(2.17) $\quad \sigma(z_j) \neq 0 \qquad (j = 1, \ldots, r)$

so that, from (2.15) with $i = j$, $\quad \rho(z_i)\,\overline{\sigma(z_i)} = -\lambda\,|\,\sigma(z_i)\,|^2 < 0$: (1.6) then implies

(2.18) $\quad |\,z_j| < 1 \qquad (j = 1, \ldots, r)$

Because of (2.18), the factor $1/(1 - z_j\,\overline{z_i})$ in (2.16) can be expanded in $\sum_{m=0}^{\infty} z_j^m\,\overline{z}_i^{\,m}$; then, from (2.16), for all column vectors $\xi \equiv (\xi_i)_{j=1,..r} \in \mathbb{C}^r$, we have:

$$\xi^*\,M\,\xi = \sum_{j;i=1}^{r}\ \sum_{m=0}^{\infty} z_j^m\,\overline{z}_i^{\,m}\,[q(z_j)\,\overline{q(z_i)} + 2\,\lambda\,\sigma(z_j)\,\overline{\sigma(z_i)}]\,\xi_i\,\overline{\xi_j} =$$

$$\sum_{m=0}^{\infty}\{(\sum_{j=1}^{r} z_j^m\,q(z_j)\,\xi_j\,)(\sum_{i=1}^{r} \overline{z}_i^{\,m}\,\overline{q(z_i)\,\xi_i}) + 2\,\lambda(\sum_{j=1}^{r} z_j^m\,\sigma(z_j)\,\xi_j)(\sum_{i=1}^{r} \overline{z}_i^{\,m}\,\overline{\sigma(z_i)\,\xi_i})\} =$$

$$= \sum_{m=0}^{\infty}\{|\sum_{j=1}^{r} z_j^m\,q(z_j)\,\xi_j\,|^2 + 2\,\lambda\,|\sum_{j=1}^{r} z_j^m\,\sigma(z_j)\,\xi_j\,|^2\}$$

so that (2.14) follows. ■

We are now ready to prove (1.8):

THEOREM 2.1. *Let* ρ, σ *be given with* (1.6); *then* (1.4) *holds true; and under the assumption* (1.7), *we have* (1.5).

Proof. Let us first assume both (1.6), (1.7). From (2.10), (2.12), we have a representation formula for the elements $g_{j,i}$ of G:

$$(2.19)\qquad g_{j,i} = \sum_{m=1}^{r} p_{j,m} \cdot \overline{p_{i,m}}$$

where $P \equiv (p_{j,m})_{j,m=1,..,r}$ is the square root of G.

Setting

$$(2.20)\qquad p_m(z) = \sum_{j=1}^{r} p_{j,m}\, z^{j-1} \qquad (m = 1, \dots, r)$$

we have $\deg(p_m) \le r - 1$; (2.11) with w replaced by $\overline{w}$ gives (1.4) ; and (1.5) follows from (2.12) (which implies P non singular).

In the case that (1.7) is not true let τ be a greatest common factor of ρ, σ; the couple $(\tilde{\rho}, \tilde{\sigma})$ with $\tilde{\rho} = \rho/\tau$, $\tilde{\sigma} = \sigma/\tau$ will admit a decomposition like (1.4) , with $\deg(\tilde{q}) \le \tilde{r}$, $\deg(\tilde{p}_m) \le \tilde{r} - 1$ $(m = 1, 2, \dots \tilde{r})$ where $\tilde{r} = r - \deg(\tau)$. Setting $q = \tau \cdot \tilde{q}$, $p_m = \tilde{p}_m \cdot \tau$ for $m = 1, \dots, \tilde{r}$ and $p_m \equiv 0$ for $m = \tilde{r} + 1, \dots, r$, we get (1.4). ■

REMARK 2.3. Starting with real α_j, β_j in (1.2) and choosing q real (see Remark 2.1), matrix G in (2.10) will be real symmetric, and polynomials p_m in (2.20) will have real coefficients.

REMARK 2.4. Using notations (2.1), (2.20), by identification of the coefficient of $z^j \overline{w}^i$ in (1.4) we get [(3)] :

$$(2.21)\quad \alpha_j \overline{\beta_i} + \overline{\alpha_i} \beta_i + \sum_{m=1}^{r} (p_{j+1,m}\, p_{i+1,m} - p_{j,m}\, \overline{p_{i,m}}) = q_j\, \overline{q_i} \qquad (j, i = 0, \dots, r)$$

which is of course equivalent (and not only a consequence) of (1.4).

Let now H be given with:

(2.22) H *is a complex Hilbert space;* (,) *and* | | *will denote the scalar product and the norm in* H

Let us define a quadratic form on H^r by setting:

---

(3) We adopt, for easier notation, the convention $p_{r,m} = 0$ $(m = 1, 2, \dots r)$; similarly, if in (2.1) it is $s < r$, we put $q_j = 0$ for $j = s+1, \dots, r$

$$(2.23)\qquad \mathcal{G}(w_1, w_2, ..., w_r) = \sum_{m=1}^{r} \Big| \sum_{j=1}^{r} p_{j,m}\, w_j \Big|^2 \qquad (w_1, ..., w_r \in H)$$

LEMMA 2.3. *Let* ρ, σ, H *be given with* (1.6), (2.22); *and let* $\mathcal{G}$ *be defined by* (2.23), *through* (2.10), (2.19). *Then:*

$$(2.24)\qquad 2\,\mathrm{Re}\,\Big(\sum_{j=0}^{r} \alpha_j w_j, \sum_{i=0}^{r} \beta_i w_i \Big) \geq \mathcal{G}(w_1, w_2, ..., w_r) - \mathcal{G}(w_0, w_1, ..., w_{r-1})$$

for all $r+1$ *- tuples* $w_0, w_1, ..., w_r$ *in* H. *Furthermore:*

$$(2.25)\qquad \mathcal{G}(w_1, w_2, ..., w_r) \geq 0 \qquad \forall\, w_1, w_2, ..., w_r \in H$$

(2.26) $\mathcal{G}$ *is strictly positive definite on* H^r *if* (1.7) *holds true*

Proof. Formula (2.23) clearly implies (2.25); from (2.12) (which implies P strictly positive in (2.19)) follows (2.26).

Expanding the left hand member of (2.23), and using (2.21) (see also [(3)]) we get:

$$2\,\mathrm{Re}\,\Big(\sum_{j=0}^{r} \alpha_j w_j, \sum_{i=0}^{r} \beta_i w_i \Big) = \sum_{j;i=0}^{r} (w_j, w_i)\,[\alpha_j \bar{\beta}_i + \bar{\alpha}_i \beta_j] =$$
$$= \sum_{j;i=0}^{m} (w_j, w_i)\,\Big[q_j \bar{q}_i + \sum_{m=1}^{r} p_{j,m} \overline{p_{i,m}} - \sum_{m=1}^{r} p_{j+1,m} \overline{p_{i+1,m}}\Big] =$$
$$= \Big| \sum_{j=0}^{r} q_j w_j \Big|^2 + \mathcal{G}(w_1, w_2, ..., w_r) - \mathcal{G}(w_0, w_1, ..., w_{r-1}) \qquad \blacksquare$$

REMARK 2.5. In fact we have obtained a more precise relation than (2.24); it is:

$$2\,\mathrm{Re}\,\Big(\sum_{j=0}^{r} \alpha_j w_j, \sum_{i=0}^{r} \beta_i w_i \Big) - \mathcal{G}(w_1, w_2, ..., w_r) + \mathcal{G}(w_0, w_1, ..., w_{r-1}) =$$
$$= \mathcal{D}(w_0, w_1, ..., w_r)$$

where $\mathcal{D}(w_0, w_1, ..., w_r) = \Big| \sum_{j=0}^{r} q_j w_j \Big|^2$.

The term $\mathcal{D}$ cannot give any help in stability estimates: however it is an important term which, in some sense, measures the "numerical dissipation" of the scheme (ρ, σ).

REMARK 2.6. Choosing in (2.22) $H \equiv \mathbb{C}$, the quadratic form $\mathcal{G}$, because of (2.19), takes the form

(2.27) $$\mathcal{G}(w_1, w_2, \ldots, w_r) = (w_1, w_2, \ldots, w_r) \cdot G \cdot (w_1, w_2, \ldots, w_r)^*$$

where V^* denotes the adjoint (column) vector of the row-vector V.

In order to prove (1.9) let us first recall that, following Dahlquist [4], a couple (ρ, σ) is G-stable if there exists a positive definite $(r \times r)$ matrix G such that, for any choice of $w_0, w_1, \ldots, w_r \in \mathbb{C}$, the following implication holds true:

(2.28) $$\begin{aligned} &\mathrm{Re}\left[\sum_{j=0}^{r} \alpha_j w_j \cdot \overline{\sum_{i=0}^{r} \beta_i w_i} \right] \le 0 \quad \Rightarrow \\ &(w_1, w_2, \ldots, w_r) \cdot G \cdot (w_1, w_2, \ldots, w_r)^* \le \\ &\le (w_0, w_1, \ldots, w_{r-1}) \cdot G \cdot (w_0, w_1, \ldots, w_{r-1})^* \end{aligned}$$

THEOREM 2.2. A-*stability and* G-*stability are equivalent.*

Proof. We confine ourselves to the "difficult" part:A-stability ⇒ G-stability,which by now is very easy: (2.24), (2.26) give (2.28), and we end up by Lemma 2.2 (or (2.27)). ■

As another application of (1.8) we gave in [2] a very short proof of the "second Dahlquist barrier":

(2.29) *The order of accuracy of* A-*stable methods can not exceed 2; for a second order* A-*stable method the best possible error constant is 1/12.*

Here we will show a different type of consequences of (1.8), concerning stability problems for the discretization (1.3) of (1.1). We will follow the scheme outlined in [1] for the so called "ϑ-method" corresponding to:

(2.30) $$\rho(z) = z - 1 \;\; ; \;\; \sigma(z) = \vartheta z + 1 - \vartheta \qquad (z \in \mathbb{C})$$

for which the decomposition (1.4) can be (for $\vartheta \geq 1/2$, as required by (1.6)) directly obtained. From (1.8) the same scheme, as we will show, applies to *any* A-stable couple ρ, σ.

3. STABILITY IN LINEAR ABSTRACT DIFFERENTIAL EQUATIONS.

Let us briefly recall the abstract setting in which we will work; for any further detail we refer to [6] Chapter 3.

Besides H, given with (2.22), let V be given with:

(3.1) V *is a complex Hilbert space;* $V \subset H$; V *dense in* H

We will denote by $\| \ \|$ the norm of V; and by V^* the completion of H with respect to the dual norm $\| h \|_* = \sup\{ |(h, v)| ; v \in V, \| v \| \leq 1\}$. It turns out that V^* is the conjugate space of V [(4)]; and there is no danger of confusion in using the same notation (,) both for the scalar product in H and for the pairing between V^* and V.

Also let

(3.2) A *be a linear continuous operator from* V *to* V^* *such that there exists* $\alpha > 0$ *with* $\mathrm{Re}(Av, v) \geq \alpha \| v \|^2 \ \forall\, v \in V$.

If f is given with:

(3.3) $f \in L^2(0, +\infty; V^*)$ [(5)]

it has a meaning to ask for u with:

(3.4) $u \in L^2(0, +\infty; V)$ [(6)]

---

(4) i.e. the space of functionals $f : V \to \mathbb{C}$ which are antilinear and continuous. Remark that in [6] such space is denoted by V', instead of V^*.

(5) Say f is a function defined (a.e.) in $(0, +\infty)$, with values in V^*, measurable and such that

$$\int_0^{+\infty} \| f(t) \|_*^2 \, dt < +\infty$$

(6) space defined in a way similar to (5); remark that, from $V \subset V^*$, such a u can be viewed as a function valued in V^*.

satisfying (1.1). We will first impose u' + Au = f is the sense of distribution with values in V*; then from u' = f - Au and (3.3), (3.4), (3.2) we get

$$(3.5) \qquad u' \in L^2(0, +\infty; V^*);$$

and finally from u' + Au = f in the distributional sense we get (1.1) a.e. in $t \geq 0$. Recall that, from (3.4), (3.5), it follows $u \in C^0([0, +\infty[; H)$, so that the following problem has a meaning:

PROBLEM 3.1. *Let* U_0 *be given with*

$$(3.6) \qquad U_0 \in H$$

and let f be given with (3.3). *We ask for u satisfying* (3.4), (1.1), *and:*

$$(3.7) \qquad u(0) = U_0$$

It is a well-posed problem: for any given $\{U_0, f\}$ with (3.6), (3.3), there exists a unique solution u,which besides satisfies the "energy identity":

$$(3.8) \quad \frac{1}{2} | u(T) |^2 + \mathrm{Re} \int_0^T (Au(t), u(t))\, dt = \frac{1}{2} | U_0 |^2 + \mathrm{Re} \int_0^T (f(t), u(t))\, dt$$

(take in (1.1), a.e. in $(0, +\infty)$, the pairing with u(t); taking the real part and by integration on (0, T), we get (3.8) as a consequence of (3.7) and the formula $2\,\mathrm{Re}\,(u'(t), u(t)) = \frac{d}{dt}(| u(t) |^2)$). From (3.2), (3.8) we also get:

$$| u(T) |^2 + 2\alpha \int_0^T \| u(t) \|^2 dt \leq | u(T) |^2 + 2\,\mathrm{Re} \int_0^T (Au(t), u(t))\, dt =$$

$$= | U_0 |^2 + 2\,\mathrm{Re} \int_0^T (f(t), u(t))\, dt \leq | U_0 |^2 + 2 \int_0^T \| f(t) \|_* \| u(t) \|\, dt \leq$$

$$\leq | U_0 |^2 + \frac{1}{\alpha} \int_0^T \| f(t) \|_*^2\, dt + \alpha \int_0^T \| u(t) \|^2\, dt.$$

Subtracting $\alpha \int_0^T \| u(t) \|^2 \, dt$ from both sides and replacing $\int_0^T \| f(t) \|_*^2 \, dt$ by $\| f \|^2_{L^2(0,+\infty;\, V^*)}$, we find that the quantity $| U_0 |^2 + (1/\alpha) \| f \|^2_{L^2(0,\, +\infty;\, V^*)}$ is an upper bound both for $| u(T) |^2$ and $\alpha \int_0^T \| u(t) \|^2 \, dt$ valid for all T, leading to:

(3.9) $\qquad \| u \|_{L^\infty(0,\, +\infty;\, H)} + \| u \|_{L^2(0,\, +\infty;\, V)} \leq C_\alpha \{ | U_0 | + \| f \|_{L^2(0,+\infty;\, V^*)} \}$

Let us now consider the discrete counterpart of Problem 3.1; ρ, σ satisfying (1.2), (1.6), (1.7). Fix any $k > 0$ and consider:

PROBLEM 3.2. *Let* $\{u_j\}_{j=0,\dots,r-1}$, $\{f_n\}_{n\geq 0}$ *be given with:*

(3.10) $\qquad u_j \in V \ (j = 0, 1, \dots, r-1)$ [7]; $\ f_n \in V^* \quad (n \geq 0)$

We ask for $\{u_{n+r}\}_{n\geq 0}$ *with*

(3.11) $\qquad u_{n+r} \in V; \ \frac{1}{k} \sum_{j=0}^{r} \alpha_j u_{n+j} + A \sum_{j=0}^{r} \beta_j u_{n+j} = \sum_{j=0}^{r} \beta_j f_{n+j} \quad$ *in* $V^* \ (n \geq 0)$

From (1.6) written for large $| z |$ we get $\operatorname{Re} \alpha_r \overline{\beta_r} > 0$; so that, by the Lax-Milgram lemma, problem (3.11) in u_{n+r} has a unique solution once $u_n, u_{n+1}, \dots, u_{n+r-1}$ are given (in addition to $\{f_n\}$); starting from (3.10), we then have existence and uniqueness for $u_r, u_{r+1}, \dots$. The whole sequence $\{u_n\}_{n\geq 0}$ is so defined; we would an estimate on it similar to (3.9)

THEOREM 3.1. *There exists a constant* C, *which depends on* α, ρ, σ *(but not on* k, H, V, A) *such that, for any choice of the data in* (3.10):

(3.12) $\qquad \sup_{n\geq 0} | u_n |^2 + k \sum_{n=0}^{\infty} \| \sum_{j=0}^{r} \beta_j u_{n+j} \|^2 \leq C \{ \max_{0\leq j\leq r-1} | u_j |^2 + k \sum_{n=0}^{\infty} \| f_n \|_*^2 \}$

Proof. Let us take the pairing of (3.11) with $\sum_{i=0}^{r} \beta_i u_{n+i}$; we get:

$$\frac{1}{k} \Big(\sum_{j=0}^{r} \alpha_j u_{n+j} , \sum_{i=0}^{r} \beta_i u_{n+i} \Big) + \Big(A \sum_{j=0}^{r} \beta_j u_{n+j}, \sum_{i=0}^{r} \beta_i u_{n+i} \Big) =$$

---

[7] for "backward methods" (i.e. $\sigma(z) = z^r$) it is sufficient to give $u_0, u_1, \dots, u_{r-1} \in H$; and $f_0, f_1, \dots, f_{r-1}$ are not used.

$$= \left(\sum_{j=0}^{r} \beta_j f_{n+j} , \sum_{i=0}^{r} \beta_i u_i \right)$$

so that, multiplying by 2k, and taking the real part, from (3.2) follows:

$$2 \operatorname{Re} \left(\sum_{j=0}^{r} \alpha_j u_{n+j}, \sum_{i=0}^{r} \beta_i u_{n+i} \right) + 2k\,\alpha \left\| \sum_{j=0}^{r} \beta_j u_{n+j} \right\|^2 \le$$

$$\le 2k \left\| \sum_{j=0}^{r} \beta_j f_{n+j} \right\|_* \left\| \sum_{i=0}^{r} \beta_i u_{n+i} \right\| \le \frac{k}{\alpha} \left\| \sum_{j=0}^{r} \beta_j f_{n+j} \right\|_*^2 +$$

$$+ \alpha k \left\| \sum_{i=0}^{r} \beta_i u_{n+i} \right\|^2$$

The last term cancels with the similar one on the left hand member; and the first term on the left hand member can be handled by means of (2.24); so that we get:

$$\text{(3.13)} \qquad \mathcal{G}(u_{n+1}, u_{n+2}, \ldots, u_{n+r}) + \alpha k \left\| \sum_{j=0}^{r} \beta_j u_{n+j} \right\|^2 \le$$

$$\le \mathcal{G}(u_n, u_{n+1}, \ldots, u_{n+r-1}) + \frac{k}{\alpha} \left\| \sum_{j=0}^{r} \beta_j f_{n+j} \right\|_*^2 \qquad (n \ge 0)$$

Summing up relations (3.13) for n = 0, 1, ..., m, most of the terms in $\mathcal{G}$ cancel and we get:

$$\text{(3.14)} \qquad \mathcal{G}(u_{m+1}, u_{m+2}, \ldots, u_{m+r}) + \alpha k \sum_{n=0}^{m} \left\| \sum_{j=0}^{r} \beta_j u_{n+j} \right\|^2 \le$$

$$\le \mathcal{G}(u_0, u_1, \ldots, u_{n+r-1}) + \frac{k}{\alpha} \sum_{n=0}^{m} \left\| \sum_{j=0}^{r} \beta_j f_{n+j} \right\|_*^2 \le$$

$$\le C \left\{ \max_{0 \le j \le r-1} |u_j|^2 + k \sum_{n=0}^{m+r} \|f\|_*^2 \right\} \le$$

$$\le C \left\{ \max_{0 \le j \le r-1} |u_j|^2 + k \sum_{n=0}^{\infty} \| f_n \|_*^2 \right\} \qquad (m \ge 0)$$

where C depends only on α, ρ, σ.

From (3.14), (2.25) (so independently from (1.7)) it follows, with a new value for C:

$$(3.15)\qquad k\sum_{n=0}^{\infty} \|\sum_{j=0}^{r} \beta_j u_{n+j}\|^2 \le C\{\max_{0\le j\le r-1} |u_j|^2 + k\sum_{n=0}^{\infty}\| f_n\|_*^2\}$$

Also the first term in (3.14) is bounded, for all $m \geq 0$, by the last one; if (1.7) holds true, it in turns bounds $C\cdot \max_{1\le j\le r} |u_{m+j}|^2$ (see (2.26)); so that, under the hypothesis (1.7) we also have, with a new C:

$$(3.16)\qquad \sup_{m\ge 1} |u_m|^2 \le C\{\max_{0\le j\le r-1} |u_j|^2 + k\|f_n\|_*^2\}$$

From (3.16), (3.15), we get (2.12). ■

REMARK 3.1. Let us assume that σ has no essential roots, namely "$\sigma(z) = 0 \Rightarrow |z| < 1$". Then, for suitably choosen C, we can prove that

$$\sum_{m=r}^{+\infty} \| u_m\|^2 \le C\cdot \sum_{n=0}^{+\infty} \|\sum_{j=0}^{+\infty} \beta_j u_{n+j}\|^2 ;$$

then, from (3.15), we get obviously:

$$(3.17)\qquad k\sum_{n=0}^{\infty} \| u_m\|^2 \le C\{\max_{0\le j\le r-1} |u_j|^2 + k\sum_{j=0}^{r-1} \| u_j\|^2 + k\sum_{n=0}^{+\infty}\| f_n\|_*^2\}$$

Let us now come back to problem 3.1. Choose any fixed $k > 0$ and discretize f by [8]:

$$(3.18)\qquad f_{k,n} = \frac{1}{k}\int_{nk}^{(n+1)k} f(t)\,dt \qquad\qquad (n \geq 0)$$

An easy computation gives:

[8] we use (3.18) because of the low regularity assumed for f (see (3.3)). For "smooth" f we could e.g. replace (3.18) by $f((n+1/2)k)$, and (3.19) would still hold

$$(3.19)\qquad k\sum_{n=0}^{+\infty}\| f_{k,n}\|_*^2 \le \int_0^{+\infty}\| f(t)\|_*^2\, dt.$$

Now choose any $\{u_{k,0}, u_{k,1}, \ldots, u_{k,r-1}\}$ in V[9], and let $\{u_{k,n}\}$ be the sequence solution of Problem 3.2 with respect to such initial data and $\{f_n\}$ given by (3.18). From (3.12), (3.19) we get:

$$(3.20)\qquad \sup_{n\ge 0} | u_n |^2 + k\sum_{n=0}^{\infty} \|\sum_{j=0}^{r} \beta_j u_{n+j}\|^2 \le$$

$$\le C\{ \max_{0\le j\le r-1} | u_j |^2 + \| f \|^2_{L^2(0,+\infty;\, V^*)} \}$$

In order to interpret the left hand member of (3.20), let us define the function $u_k(t)$ by means of:

(3.21) $u_k(t)$ *is the step function which, for* $n \ge 0$, *takes the value* $u_{k,n}$ for $t \in\,]n\,k, (n+1)\,k\,[$ [10] *and the function* $\tilde{u}_k(t)$ *by means of:*

(3.22) $\tilde{u}_k(t)$ *is the step function which, for* $n \ge 0$, *takes the value* $\sum_{j=0}^{r} \beta_j u_{n+j}$

Then (3.20) reads:

$$(3.23)\qquad \| u_k \|_{L^\infty(0,+\infty;\, H)} + \| \tilde{u}_k \|_{L^2(0,+\infty;\, V)} \le$$

$$\le C\{ \max_{0\le j\le r-1} | u_j | + \| f \|_{L^2(0,+\infty;\, V^*)} \}$$

Starting from (3.23), a standard argument ("consistency + stability $\Rightarrow$ convergence, and error estimates for smooth solutions") will imply that, under the usual consistency assumptions:

$$(3.24)\qquad \rho(1) = 0; \qquad \rho'(1) = \sigma(1) \ne 0$$

---

(9) See also (7). Of course we would choice $u_{kj} \cong u((j+(1/2))k)$, because we espect $u_{k,n} \cong u((n+(1/2))k)$ $\forall\, n \ge 0$; for first order methods it is sufficient to choose $u_{k,j} \equiv U_0$ (or an approximation in V of U_0); for second order methods we can use the expansion $u((j+(1/2))k)$ $u(0) + (j + (1/2))\,k\,u'(0) = U_0 + (j + (1/2))\,k\,[f(0) - A\,U_0]$ (see (3.7) and (1.1) at $t = 0$)

(10) More generally we could choose $u_k(t) = \sum_{n=0}^{+\infty} u_{k,n} \cdot (\phi\,(t/k) - n)$ with ϕ "smooth"; our choice corresponds to $\phi = \chi_{]0,1[}$. Different choices would just change the value of C in (3.23).

the errors $u(t) - u_k(t)$ (in the norm of $L^\infty(0, +\infty; H)$) and $u(t) - \tilde{u}_k(t)/\sigma(0)$ (in the norm of $L^2(0, +\infty; V)$) tend to 0 as $k \to 0^+$ if $\{u_{k,j}\}_{j=0,\dots,r-1}$ are choosen as suggested in [9]; and can be bounded (by $C\,k$ or $C\,k^2$ according to the accuracy order of the method; see (2.29)) if the data U_0, $f(t)$ are "smooth and compatible" [11].

REMARK 3.2. In the framework of Remark 3.1, by adding $k\,C \sum_{j=0}^{r-1} \| u_{k,j} \|^2$ at the right hand member of (3.23), we could replace $\| \tilde{u}_k \|_{L^2(0,\, +\infty;\, V)}$ by $\| u_k \|_{L^2(0,\, +\infty;\, V)}$ (and similarly in the error estimates). In the general case however this modified estimate can fail, as handly checked for the Crank-Nicholson scheme (corresponding to $\vartheta = 1/2$ in (2.30)).

Let us make a remark about problem 3.1: because of the density of V in H, condition (3.7) can also be imposed by asking that, for all $v \in V$:

(3.25) $$(u(0), v) = (U_0, v)$$

Directly from the definition of equality in V^*, equation (1.1) can be equivalently be written by imposing that, for all $v \in V$:

(3.26) $$(u'(t), v) + (Au(t), v) = (f(t), v) \qquad \text{a.e. in } t > 0$$

In a similarly way, the equation in (3.11) can be written by imposing, for all $v \in V$:

(3.27) $$\frac{1}{k}\Big(\sum_{j=0}^{r} \alpha_j u_{n+j}, v\Big) + \Big(A \sum_{j=0}^{r} \beta_j u_{n+j}, v\Big) = \Big(\sum_{j=0}^{r} \beta_j f_{n+j}, v\Big)$$

Let us now consider the following generalization [12] of Problem 3.1, 3.2: we will assume that a space W is given with:

(3.28) W *is a closed subspace of* V

[11] e.g. for $U_0 \in V$, $f(t), f'(t) \in L^2(0, +\infty; V^*)$, $Au_0 - f(0) \in H$, we get $u, u' \in L^2(0, +\infty; V)$, $u'' \in L^2(0, +\infty; V^*)$. The compatibility condition "$Au_0 - f(0) \in H$" cannot be dropped

[12] of course, choosing in (3.28) $W = V$, we get again Problems 3.1, 3.2.

PROBLEM 3.3. *Let* U_0, f *be given with* (3.6), (3.3). *We ask for* U *with:*

(3.29) $U \in L^2(0, +\infty; W)$ (13)

such that, for all $v \in W$, (3.25), (3.26) *hold true.*

PROBLEM 3.4. *Let* $\{u_j\}_{j=0,\dots,r-1}$, $\{f_n\}_{n\geq 0}$ *be given with* (3.10). *We ask for* $\{U_{n+r}\}_{n\geq 0}$ *with:*

(3.30) $U_{n+r} \in W \qquad (n \geq 0)$

such that for all $v \in W$, *for all* $n \geq 0$, (3.26) *holds true.*

Remarking that all estimates we obtained were obtained by taking some pairings which now are already written (it is now sufficient to choose v = U(t) in (3.26), $v = \sum_{j=0}^{r} \beta_j U_{n+j}$ in (3.27)) all the estimates like (3.9), (3.2), (3.23) still remains valid; and *all constants in such estimates do not depend on* W.

REMARK 3.3. Let us illustrate the abstract setting with a concrete example.

Consider, in a bounded open subset of Ω of $\mathbb{R}^3$, the Cauchy-Dirichlet problem for the heat diffusion equation:

$$\text{(3.31)} \qquad \begin{aligned} &\partial u/\partial t - \sum_{l=1}^{3} \partial^2 u/\partial x_l^2 = f(x,t) && x \in \Omega, \quad t > 0 \\ &u(x, t) = 0 && x, \in \partial\Omega, t > 0 \\ &u(x, 0) = U_0(x) && x \in \Omega \end{aligned}$$

Looking at u(x,t), f(x,t) as functions of t with values in spaces of functions of x, we will choose $H = L^2(\Omega)$ in (2.22). Denoting by $H^1(\Omega)$ the usual Sobolev space $\{v \in L^2(\Omega) \mid \nabla v \in \{L^2(\Omega)\}^3\}$, we will choose in (3.1) $V = H_0^1(\Omega) =$ $= \{v \in H^1(\Omega) \mid v_{|\partial\Omega} = 0\}$; and in (3.2) $A = \sum_{l=1}^{3} \partial^2/\partial x_l^2$. Then Problem 3.1 coincides with (3.31), and Problem 3.3 with $W = H_0^1(\Omega)$ (and (Au, v) replaced by

(13) W is endowed with the norm of V

$\int_\Omega \nabla u \cdot \nabla v \, dx$) is the usual weak formulation of (3.31).

A Faedo Galerkine approximation of (3.31) is obtained simply by solving Problem 3.3 where the space W in (3.28) is a (finite dimensional) subspace of $\overset{1}{H_0}(\Omega)$, given e.g. by trigonometric polynomials if Ω is rectangular, or by finite elements on Ω, or any other subspace we prefer. In general, if W= span $\{v_1, v_2, ..., v_N\}$, the solution U of Problem 3.3 will have the form $U(x, t) = \sum_{n=1}^{N} \xi_n(t)\, v_n(x)$; and Problem 3.3 can of course by written as a system of ordinary differential equations in the unknown column vector $\xi(t) \equiv (\xi_n(t))_{n=1, ..., N}$; e.g. if $\{v_j\}_{j=1,...,N}$ are choosen orthonormal in $L^2(\Omega)$, Problem 3.3 reads:

(3.32) $\qquad \xi'(t) + \mathcal{A}\,\xi(t) = \phi(t) \qquad (t > 0); \quad \xi(0) = \xi_0$

for suitable matrix $\mathcal{A}$, and ξ_0, $\phi(t)$; Problem 3.4 is then the (ρ, σ) discretization of (3.32).

4. ASYMPTOTIC EXPANSION AND EXTRAPOLATION.

In order to avoid the technical difficulties outlined in [11], we will confine ourselves to U_0, f given with:

(4.1) $\qquad U_0 = 0;$ *for some* $t_0 > 0 \quad f|_{(0,t_0)} \equiv 0$

In fact, when (4.1) holds, for any $s \geq 0$ $f \in H^s(0, +\infty; V^*)$ the solution will belong to $H^s(0, +\infty; V) \cap H^{s+1}(0, +\infty; V^*)$ (both in Problem 3.1 and in Problem 3.3).

For $u_k(t)$ given by (3.21) (or better: constructed as in [11] with a C^∞ function ϕ) we could expect, for infinitely smooth f and u, an asymptotic expansion like [14]:

(4.2) $\qquad u_k(t) \cong v_0(t) + k\, v_1(t) + ... + k^n\, v_n(t) + ...$

where, of course, $v_0 \equiv u$; and $v_1, v_2, ..., v_n, ...$ can be easily characterized because of the linearity of problem 3.1; e.g. we espect v_1 to be the solution of:

---

[14] we assume, of course, that $u_0, u_1, ..., u_{r-1}$ are all 0; see [9]

$$(4.3) \qquad v'_1(t) + Av_1(t) = \frac{2\sum_{j=0}^{r} j\,\beta_j - \sum_{j=0}^{r} j^2\,\alpha_j}{2\,\sigma(1)}\, u''(t)); \qquad v_1(0) = 0$$

In order to avoid technical difficulties connected with (3.22), let us work in the framework of Remark 3.1. By the same argument used in section 3 we can prove that the v_1 given by (4.3) satisfies:

$$(4.4) \qquad \| u_k - u - k\, v_1 \|_{L^2(0,\infty;\, V) \cap L^\infty(0,\infty;\, H)} = 0(k^2)$$

if f is given with $f \in H^2(0, +\infty; V^*)$; more generally we can exhibit an explicit formula (which, as in (4.3), requires to solve some problems) for $v_2, v_3, ..., v_n, ...$ in (4.2), such that, with $v_0 = u$:

$$(4.5) \qquad \| u_k - \sum_{l=0}^{n} k^l\, v_l \|_{L^2(0,\, +\infty;\, V) \cap L^\infty(0,\, +\infty;\, H)} = 0(k^{n+1})$$

if $f \in H^{n+1}(0, +\infty; V^*)$.

REMARK 4.1. Achievement of (4.4), (4.5) just requires the evaluation of some pairings; the same proof will work for problems 3.2, 3.4, with a formula like:

$$U_k(t) \cong V_0 + k\, V_1 + ... + k^n\, V_n + ...$$

where $V_0 = U$ and e.g. V_1 is given by:

$$V_1 \in L^2(0, +\infty; W); \quad \forall\, v \in W \;\; (V'_1(t), v) + (A\, V_1(t), v) =$$

$$= \Big(\frac{2\sum_{j=0}^{r} j\,\beta_j - \sum_{j=0}^{r} j^2\,\alpha_j}{2\sigma(1)}\, U'', v\Big) \;\; : \;\; V_1(0) = 0$$

and in the analog of (4.4), (4.5), we will have:

(4.6) $\qquad 0(k^{n+1})$ *is uniform with respect to* W

Let us point out the interest of formulas like (4.4), (4.5): *independently of the knowledge of* v_1, let us solve Problem 3.2 with respect to some k, to get a first approximation [15] u_k for u; then solve Problem 3.2 with k replaced by k/2, getting a second approximation $u_{k/2}$. Defining

---

(15) in general of first order; see later for second order methods

$$(4.7) \qquad \overset{*}{u_k} = 2\, u_{k/2} - u_k$$

we will have:

$$\overset{*}{u_k} - u = 2(u_{k/2} - u - (k/2)\, v_1) - (u_k - u - k\, v_1)$$

Booth terms being $0(k^2)$ we get

$$(4.8) \qquad \| \overset{*}{u_k} - u \|_{L^2(0, +\infty; V) \cap L^\infty(0, +\infty; H)} = 0(k^2)$$

Of course, for second order methods, the coefficients of u" on the right hand member of (4.3) vanishes, so $v_1 \equiv 0$. Formula (4.7) will then be replaced by:

$$(4.9) \qquad \overset{*}{u_k} = \frac{4}{3}\, u_{k/2} - \frac{1}{3}\, u_k$$

and, instead of (4.8), we will have:

$$(4.10) \qquad \| \overset{*}{u_k} - u \|_{L^2(0, +\infty; V) \cap L^\infty(0, +\infty; H)} = 0(k^3)$$ (16)

REMARK 4.2. This was the well known "Richardson extrapolation argument", already suggested in [3] in order to overcome the lack of accuracy due to the second barrier (see (2.29)). From the proof of [3], however, it is not clear if, in the framework of Remark 4.1, (4.6) holds true; our proof (which, unfortunately, work only for linear problems) gives automatically (4.6).

Let us end up with a "computational" remark. The Richardon extrapolation is very poorly adapted to parallel computations: if one processor solves the problem in u_k (or better: in U_k, which is the true problem one can handle by a computer!) and another one solves the problem for k/2, the first processor will have nothing to do half the time!

A better exploitation of the computational resources will be obtained by the following trick that we will describe just for the ϑ-method (see (2.30)), but which works in general. Fix ϑ_1, ϑ_2 both greater then 1/2 and, *with the same* k, solve Problem 3.4 obtaining, with obvious notations, two approximations U_{k,ϑ_1} and

---

(16) by using (4.5) with n = 3, if also v_3 vanishes we will end up with an $0(k^4)$; this is the case for the Crank Nicholson scheme, where all v_{2n+1} in (4.2) vanish.

U_{k,ϑ_2} for U (and the evaluations of U_{k,ϑ_1}, U_{k,ϑ_2} will require the same amount of time). Let V be the solution of

$$V \in L^2(0, +\infty; \vartheta); \quad (V'(t), v) + (AV(t), v) = (1/2) \; (U''(t), v) \; \forall v \in \vartheta; \quad V(0) = 0.$$

By (4.3) we have, for j = 1, 2:

$$U_{k,\vartheta_j} - U - k(2\vartheta_j - 1)V = 0(k^2)$$

so that, setting $U_k^* = \dfrac{(2\vartheta_2 - 1)U_{k,\vartheta_1} - (2\vartheta_1 - 1)U_{k,\vartheta_2}}{2(\vartheta_2 - \vartheta_1)}$,

we have $U_k^* - U = 0(k^2)$.

REFERENCES.

[1] C.BAIOCCHI. Stabilité uniforme et correcteurs dans la discrétisation des problemes paraboliques. In Research Notes in Mathematics, 70 (50-67) Pitman, London, 1982.

[2] C.BAIOCCHI, M.CROUZEIX. On the equivalence of A-stability and G-stability. To appear on Applied Numerical Mathematics. Special Issue.

[3] G.DAHLQUIST. A special stability problem for linear multistep methods. BIT, 3 (1963) 27-43.

[4] G.DAHLQUIST. Error analysis for a class of methods for stiff non linear initial value problems. Lecture Notes in Mathematics, 506 (60-72) 1976.

[5] G.DAHLQUIST. G-stability is equivalent to A-stability. BIT, 21 (1978) 384-401.

[6] J.L.LIONS, E.MAGENES. Non homogeneous boundary value problems and applications. T.1. Grund. Math. Wiss., 181 (1972) Springer, Berlin.

PARALLELISM ACROSS THE STEPS FOR DIFFERENCE AND DIFFERENTIAL EQUATIONS

A.BELLEN(*)
Dipartimento di Scienze Matematiche
Università di Trieste
34100 Trieste - Italia

1. INTRODUCTION.

Initial value problems such as:

$$z_{n+1} = F_{n+1}(z_n)$$

(1)

$$z_o \quad \text{given}$$

where $z_n \in R^m$, $F_n : R^m \to R^m \; \forall n$, are typically sequential problems since, by definition, the vector z_{n+1} can be obtained only by the knowledge of z_n and this holds for all $n = 0, 1, \ldots$

Equation (1) models discrete dynamical systems including, as a special case, solutions of ode initial value problem such as

(1')
$$y'(t) = f(t, y(t))$$
$$y(t_o) = y_o$$

on discrete sets of points $t_o < t_1 < \ldots t_n$.

Denote them by

$$y_{n+1} = y(t_{n+1}, t_n, y_n)$$

where $y(t_{n+1}, t_n, y_n)$ is the solution of

(*) This work was supported by M.P.I. and C.N.R.

$$y'(t) = f(t, y(t))$$
$$y(t_n) = y_n$$

at the point t_{n+1}, or an approximation of it by some, possibly implicit, one step numerical method.

A first approach, towards an implementation of (1) on a multiprocessor machine, is to expose any possible concurrency in the computation of each F_n. This concurrency could arise either from partitioning the original problem in subproblems, or from intrinsic parallelism of each individual component of the problem itself (for example uncoupled stages of a Runge-Kutta method).

An overview of these kinds of parallelism, called "across the system" and "across the method" for numerical ode's initial value problems, as well as a critical presentation of various concrete parallel approaches are exhaustively surveyed by Gear in [3] and [4]. Recent papers on this line are I. Lie [9], Jackson-Nørsett [8] Nørsett-Simonsen [12], van der Houwen-Sommeijer [6], van der Houwen-Sommeijer-van Mourik [7].

Despite the intrinsically sequential character of (1'), a sort of parallelism can be detected also "across the time" for both small scale and large scale parallelism. This paper addresses the case of large scale parallelism from a purely algorithmic point of view, that is, disregarding all the implementation devices related to the particular computational environment (hardware and software) in which the algorithm is supposed to run.

For equation (1'), consider the Picard iteration

$$dy^{(k)}/dt = f(t, y^{(k-1)}) \qquad y^{(0)}(t) = y_0$$

whose solution is

$$y^{(k)}(t) = y_0 + \int_{t_0}^{t} f(s, y^{(k-1)})ds$$

The use of quadrature rules on a mesh $t_0, \ldots t_N$ for the approximation of the integrals permits a high degree of parallelism because values $f(t_i, u^{(k-1)}(t_i))$ can be computed concurrently for all i = 1, ..., N. The values $y^{(k)}(t_i)$ need weighted sums and all can be done with parallel computational complexity $\log_2 N$. In case of a very large integration interval as well as of highly accurate solutions, the method is supposed to be applied on a series of windows depending of the number N of processors available.

Although iterates converge to the solution y for any Lipschitz function f, the convergence might be too slow. Faster iterations can be obtained by modifications of the Picard iteration such as the "shifted Picard"

$$dy^{(k)}/dt - Ay^{(k)} = f(t, y^{(k-1)}) - Ay^{(k-1)} \qquad A = \text{constant}$$

or the so-called "waveform relaxation" methods such as

$$dy^{(k)}/dt - J(y^{(k-1)})\, y^{(k)} = f(t, y^{(k-1)}) - J(y^{(k-1)})\, y^{(k-1)}$$

where J(y) is the Jacobian matrix $\partial f/\partial y$ (Newton-Picard) or its lower triangular part (Gauss-Seidal-Picard) or its diagonal (Jacobi-Picard). These methods need, each iteration, the integration of a linear system which can be done in parallel across the time by means of the superposition principle, with parallel complexity $\log_2 N$. In particular the wareform Jacobi method has the nice feature that the system has independent rows and hence parallelism across the system can also be exploited.

In principle, iterative methods have a great potential of parallelism and their performances depend on the convergence rate of iterations which is in turn influenced by the stiffness of the equation. As pointed out by Gear in [5], any ODE lies between two extremal cases, the infinitely non-stiff problem $y' = f(t)$ and the infinitely stiff-problem $f(y, t)=0$. They can be solved, at N mesh points, with parallel complexity $\log_2 N$ and finite respectively. Starting from this observation, he proposes a parallel version of a blended stiff-non stiff method, inspired to the Skeel-Kong approach [13], with parallel complexity $\log_2 N$ and supplied by convergence and stability analysis. All these approaches are based on solving, by appropriate numerical methods, the sequence of modified Picard iterations.

A different approach is to choose first the appropriate discretization method for the initial value problem and to define the recursion (1) as its sequential implementation, second to approximate the solution of (1) by some iterative method exploiting parallelism across the indeces, i.e. "across the steps".

Chapter 2 and 3, based on [1], develop the error analysis and practical implementation of a class of parallel iterative methods for the general difference equation (1). Chapter 4 deals with the specific case of ODEs to be solved on a prefixed mesh and outlines some implementation hurdles inherent in this setting.

Two possible strategies for the parallel implementation of a step size varying method based on the previous theory are described and analyzed from a plurely speculative point of view.

2. PARALLELISM ACROSS THE STEPS.

View the sequence $\mathbf{z} := \{z_0, z_1, \ldots\}$, determined by (1), as the fixed point of the following transformation Φ acting on the set S of sequences $\mathbf{u}$ with first component $u_0 = z_0$:

$$\Phi(\mathbf{u}) := \{u_0,\ F_1(u_0),\ F_2(u_1), \ldots F_n(u_{n-1}), \ldots\}.$$

If $\mathbf{u}^0 \in S$, is an arbitrary initial sequence, the recurrence

$$\text{(2)} \qquad \mathbf{u}^{k+1} = \Phi(\mathbf{u}^k) \qquad k = 0, 1 \ldots$$

evidently converges to the solution $\mathbf{z}$, since at the i-th iteration $u_j^i = z_j$ for all $j \le i$.

Under the assumption that the concurrent computation of the right hand side of (2) takes the same time needed for the computation of one step of the recurrence (1), the number of iterations of (2) must be significantly less than the number of steps to be processed in order to achieve any parallel speed up. This depends on the convergence rate of (2). In order to increase this rate consider the following improved iteration:

$$\text{(3)} \qquad \mathbf{u}^{k+1} - D(\mathbf{u}^k)\ \mathbf{u}^{k+1} = \Phi(\mathbf{u}^k) - D(\mathbf{u}^k)\,\mathbf{u}^k$$

where $D(\mathbf{u}^k)$ is a subdiagonal m-block matrix

$$D(\mathbf{u}^k) = \mathrm{subdiag}(\Gamma_1(\mathbf{u}^k), \Gamma_2(\mathbf{u}^k), \ldots) \qquad \Gamma_i \in \mathcal{L}(R^m, R^m).$$

For instance, the Newton method provides $D(\mathbf{u}^k) = \Phi'(\mathbf{u}^k)$ and hence $\Gamma_i(\mathbf{u}^k) = F'_i(u_{i-1}^k)$ while the modified Newton method provides $\Gamma_i(\mathbf{u}^k) = F'_i(u_{i-1}^o)$. The Steffensen method provides, in scalar case:

$$\Gamma_i(\mathbf{u}^k) = \lambda_i(\mathbf{u}^k) = \frac{F_i(F_{i-1}(u_{i-2}^k)) - F_i(u_{i-1}^k)}{F_{i-1}(u_{i-2}^k) - u_{i-1}^k}$$

while, in vector case, the matrices $\Gamma_i(\mathbf{u}^k)$ take a more complicated form (see [1]). In any case we shall consider iterations (3) fulfilling the following "*local convergence property* ". For the sequence $\mathbf{u}^k$, define the error:

$$e_i^k := z_i - u_i^k \qquad i = 0, 1, ...$$

and the *maximum error*

$$E_n^k = \max_{\nu \le n} \{e_\nu^k\}.$$

LOCAL CONVERGENCE PROPERTY: *There exist an integer* p *and constants* $C_{n+1} > 0$ ($C_{n+1} < 1$ if $p = 1$) *such that:*

(4) $$E_{n+1}^{k+1} \le C_{n+1} (E_n^k)^p \qquad \forall_{k \ge 0}$$

for every initial sequence $\mathbf{u}^0$ *in a suitable neighborhood of the solution* $\mathbf{z}$.

Since both $\Phi(\mathbf{u}^k)$ and $D(\mathbf{u}^k)$ in (3) depend on $\mathbf{u}^k$ they can be computed concurrently each iteration. On the other hand the n-th component is

$$u_n^{k+1} - \Gamma_n(\mathbf{u}^k)\, u_{n-1}^{k+1} = F_n(u_{n-1}^k) - \Gamma_n(\mathbf{u}^k)\, u_{n-1}^k$$

and must be computed recursively by the recurrent formula:

(5) $$u_n^{k+1} = \Gamma_n(\mathbf{u}^k)\, (u_{n-1}^{k+1} - u_{n-1}^k) + F_n(u_{n-1}^k) \qquad n = 1, 2, ...$$

Summarizing, the parallel algorithm for the non linear recurrence (1) consists in iterations of still recurrent but affine equations (5). No matter which the functions F_n and the matrices $\Gamma_n(\mathbf{u}^k)$ are, each step of the recurrent formula (5) takes the standard form of an affine operator in R^m. The time needed for the parallel computation of each individual affine operator will be referred to as the "time unit" u.

The recurrence (5) represents the unavoidable sequential part of (1) and its sequential implementation on the range n = 1, ... n* should require n* time units each iteration. Nevertheless the number of time units can be reduced to log(n*) by a parallel implementation of (5) (see Schendel [11]). Even in this case, for a large n*, the recurrence (5) determines the most time consuming part of each iteration.

One can guess that, in principle, the lower the number of iterations and the larger the computational complexity of F_n's and Γ_n's, the larger the attainable speed up. Before formalizing and motivating better the claim, some remarks and comments are necessary about the concept of speed up for an algorithm.

A frequent definition of speed up for an algorithm is

$$\text{speed up} = \frac{\text{execution time using one processor}}{\text{execution time using n processors}}$$

Sometimes this definition can be misleading since, as noticed by various authors, it exhibits the parallel character of the algorithm rather than to give a measure of its performance. Therefore it seems better to adopt the more pragmatic definition:

$$(6) \quad \text{speed up} = \frac{\text{execution time of the fastest sequential algorithm}}{\text{execution time of the parallel algorithm}}$$

where we assume that the fastest sequential algorithm for the numerical solution of (1) is just the recurrent computation if (1) itself. Also this definition gives rise to some doubts in giving a concrete meaning to the lower term of the ratio. In fact the execution time of a parallel algorithm strongly depends on the specific architecture employed, in particular on the communication system among processors and between processors and memories, so that the base-parallel algorithm requires different implementation devices according to the targeted architecture and, in terms of execution time, the results can be quite different. As a matter of fact, the divergent development of multiprocessors does not yet allow us to talk about "portability" of an algorithm, neither with respect to implementation languages nor with respect to architectures. Therefore, in defining speed up for algorithms by the ratio (6), the computational environment, i.e. hardware and software, ought to be specified.

A second minor source of uncertainty in (6) is that we are comparing an exact algorithm with an approximate one whose accuracy depends on the number or iterations which evidently influences the execution time.

In spite of these concerns, it is worthwhile to consider a theorical assessment of speed up as independent as possible of any specific computational environment. To this aim we make the hypothesis that each individual instruction takes the same time in sequential and parallel implementation. Moreover we assume that, if a given instruction has to be executed n times, the recurrent execution time on a single processor is n time larger than the concurrent execution time on n processors. Finally for both sequential and parallel algorithms communication time is neglected in the speed up estimate. In such a way we obtain an attainable speed up evaluation which is intrinsic to the parallel algorithm. Afterwards one has to single out the computational environment suited for the best exploitment of the potential parallelism.

For our class of algorithms, assume that T is the number of time units needed for the computation of each F_n. The execution time of the recurrence (1), for $n \le n^*$, is then n^*Tu. On the other hand assume that the parallel algorithm requires s

iterations of the recurrence (5) and that, each iteration, the ingredients $F_n(u^k_{n-1})$ and $\Gamma_n(\mathbf{u}^k)$, $n \leq n^*$, are computed concurrently in νT time units. Moreover assume that the recurrence (5), for $n \leq n^*$, is computed by some parallel algorithm taking $\log(n^*)$ time units. The global execution time of the parallel algorithm is then:

$$s\,(\log(n^*) + \nu T)u$$

The choice of the iteration method, that is of the matrices $\Gamma_n(\mathbf{u}^k)$, determines the constant ν. In particular for the Steffensen method $\nu = 2$, since each matrix needs two nested computations of F.

The attainable speed up is then:

$$\text{speed up} = \frac{n^*T}{s(\log(n^*) + \nu T)} \tag{7}$$

It is worth just to mention that for linear problems, i.e. for affine functions F_n's, one iteration is sufficient for convergence to the exact solution by any iteration fulfilling the local convergence property with order $p > 1$. In this case the algorithm is nothing but the recurrent implementation of (1) after that all computations independent of $\mathbf{z}$ have been previously and concurrently carried out throughout the range $[0, n^*]$.

Formula (7) shows that, unlike algorithms exploiting parallelism across the system and across the method, in parallelism across the steps the larger T, the larger the attainable speed up on any finite range $n \leq n^*$. Moreover, in order to have a speed up bigger than one, the number of iterations cannot exceed $n^*\,T/(\log(n^*) + \nu T)$. In particular, for $s \equiv s(n^*)$ growing like $n^*/\log(n^*)$, the speed up tends to cT, where $c = \lim(n^*/s\,\log(n^*))$, while for a slower growth of s, the speed up diverges as n^* tends to infinite.

A small number of iterations, i.e. a sharp evaluation of the initial guess $\mathbf{u}^0$, is crucial for a good performance of the algorithm.

In actual implementation of the algorithm on a general purpose parallel computer, one must expect a drastic reduction in the real speed up (6) caused by communication time. On the other hand in a dynamical process control which requires a very fast (real time) response modelled by (1), a special purpose architecture could be designed in order to exploit as much as possible the parallelism intrinsic of the algorithm so as to attain a real speed up close to the best attainable speed up (7).

3. PRACTICAL IMPLEMENTATION.

In practical implementation of some algorithm based on iteration (3), one is faced with the following problems.

1- The number of processors available is, in general, less than the number of steps to be processed.
2- An initial guess $\mathbf{u}^o$, fulfilling the local convergence property, is needed to start iterations.
3- A stopping criterion is needed to stop iterations according to some error test.

Let us assume that we want to solve equation (1) for $n \leq n^*$, and we have a number of processors available allowing us to treat blocks of N $(< n^*)$ steps concurrently. If no suggestions about a reasonable initial guess $\mathbf{u}^o$ is given by the knowledge of the problem modelled by (1), or by some prediction via a low-cost, low-accuracy method, we must be satisfied with constant initial guess $u_o^o = u_1^o = u_2^o \ldots = u_N^o = z_o$.

At first sight one is tempted to proceed by iterating the first block until the stopping criterion (we suppose to have available) is fulfilled; then to process in the same way a second block, and so on. This strategy is not the most appropriate because the behavior of the iterations on each block ultimately depends on the initial guess. A rough guess on the underlying block, possibly not lying inside the neighborhood where the local convergence property holds, will cause the iteration to work like the sequential algorithm getting one step of the solution per iteration. On the other hand, even if this does not occur, it can well be that different steps of the same block need different numbers of iterations for achieving the wanted accuracy so that, during the iterations, more and more processors will lie idle.

To overcome these inconveniences one can perform a slightly more complicated but much more effective strategy allowing us to advance while iterating. It is based on the following remarks.

For any iteration fulfilling inequality (4), the maximum error E_n^k, which evidently is non-decreasing with respect to n for each k, turns out to be decreasing with respect to k. Therefore, given a positive number δ, the largest n for which $E_n^k < \delta$ is a non-decreasing function of k. This fact suggests that after each iteration the underlying block could be splitted in three segments, some of which possibly empty. A first segment S_1 including steps whose values can be accepted as good approximation of the solution, according to some error test. A second segment S_2 including steps whose values belong to the neighborhood where the local convergence property holds (essentially values which have been contracted by the iteration function Φ) which need to be re-iterated to get the wanted accuracy. A last segment S_3 of remaining steps whose values must drop and be replaced by new guesses. The iteration function is then applied to a shifted block defined by skipping S_1 and adding to the right of S_2 as many new guesses as necessary to form a new block of N steps (fig. 1).

Such a strategy allows us to advance while iterating so as to exploit each iteration all the processors available, and to guarantee for each n a convergent sequence u_n^k of actual order p.

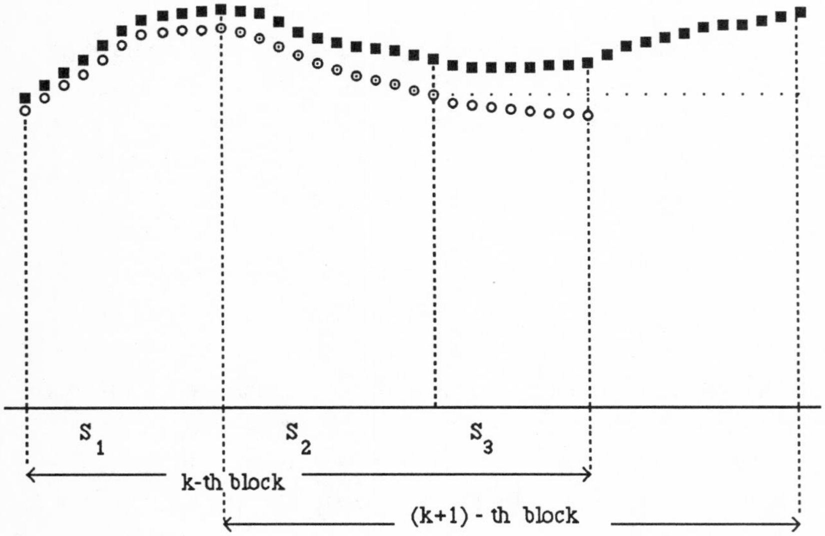

▪ true solution
∘ values given by the k-th iteration
· starting values for the $(k + 1)$ - th iteration

The last thing to be done is to provide an error test for determining, after each iteration, the segments S_1 and S_2. To this aim let us define, for $\mathbf{u}^k$, the *local error:*

$$\tau_n^k = F_n(u_{n-1}^k) - u_n^k \qquad n = 1, \ldots, N$$

and state the following theorem proved in [1].

Theorem 1. *For every* $n \geq 1$, *there exists a constant* K_n *such that, for* $\mathbf{u}^k$ *in a suitable neighborhood of* $\mathbf{z}$,

$$E_n^k \le K_n \max_{\nu \le n} \| \tau_\nu^k \|.$$

Given a tolerance TOL, the segment S_1 of the k-th block ends with the largest index s such that $\| \tau_\nu^k \| \le$ TOL for all $\nu \le s$. The maximum error E_n^k turns out to be bounded by k_s TOL in the segment [0, s], and by k_{n*} TOL in the whole range [0, n*].

Let us observe that the computation of the local errors τ_n^k along the underlying block can be done concurrently and does not add any extrawork to the algorithm since the $F_n(u_{n-1}^k)$' s are needed in the recurrence (5) for the next iteration $\mathbf{u}^{k+1}$.

As for the size of the segment S_2, it is reasonable to consider as contractive those steps j where

$$\max_{\nu \le j} \| \tau_\nu^{k+1} \| \le \max_{\nu \le j} \| \tau_\nu^k \| \tag{8}$$

The attainable speed up formula takes now the form:

$$\text{speed up} = \frac{n^*T}{s(\log(N) + \nu T)}$$

where s is the number of iterations, i.e. the number of blocks processed to reach n*, and νT is the number of time units required by the concurrent computations and by the tests for determining the segment S_1 and S_2 on each block.

Since overlapping between blocks are expected, the integer s will be larger than n*/N. If the block size, that is the number N, is too small, a large number of iterations s is needed and the speed up collapses to values less than one because the term sνT dominates. For N greater then this minimum we achieve parallel speed up until N is larger than same maximum value after which the speed up is lost. This occurs because for too large N we expect to have in each iteration large segments S_3 in which all the computations will end completely wasted.

4. APPLICATION TO ODE'S.

A natural application of the parallel algorithm presented above is numerical initial value problem for ordinary differential equations since any one-step method, on a fixed mesh $\{t_0, t_1, ... t_{n*}\}$, can be stated as a difference equation (1).

Therefore, for a given tolerance, the parallel algorithm gives an approximation $\mathbf{u}^k$ of $\mathbf{z}$, which is itself an approximation of $\mathbf{y}$, the solution of the ode-initial value problem at the mesh points. The difficulty is to give a tolerance such that the error $\| \mathbf{u}^k - \mathbf{z} \|$ is of the same magnitude as $\| \mathbf{y} - \mathbf{z} \|$, or larger, so as to avoid extra iterations leading to approximations closer and closer to $\mathbf{z}$ but not to $\mathbf{y}$.

The idea of performing a criterion for stopping iterations based on estimates of $\| \mathbf{u}^k - \mathbf{y} \|$ rather than $\| \mathbf{u}^k - \mathbf{z} \|$ takes sense only if $\| \mathbf{y} - \mathbf{z} \|$ is smaller than the wanted accuracy, that is, if $\mathbf{z}$ can be viewed as the exact solution of the ode initial value problem. Unfortunatly this depends on the mesh which is fixed and does not allow any "a priori" error estimate.

Two strategies are then possible in order to provide a parallel algorithm performing the step size control and exploiting the parallelism across the steps simultaneously.

The first strategy consists in fixing a set of equispaced points $\{t_0, t_1, ... t_{n*}\}$ in the range $[t_0, t_{n*}]$ where the solution is sought, and carring out the algorithm for equation (1) where $F_{n+1}(z_n)$ represents the numerical approximation of $y(t_{n+1},t_n,z_n)$ obtained by some variable step size method.

Given an initial guess $\mathbf{u}^0 = \{u_0^o = y_0,\ u_1^o, ...,\ u_{n*}^o\}$, the algorithm proceeds as shown in the previous sections with the sole difference that functions F_n differ from each other and therefore their concurrent computations need a MIMD machine. Moreover syncronization is needed at the end of each parallel stage.

The step size control in the computation of each $F_{n+1}(z_n)$ is based on a given discretization tolerance TOL_δ and, under suitable smoothness conditions on f, it leads to global error bounds

$$\| y(t_{n+1}, t_n, z_n) - F_{n+1}(z_n) \| \leq c\, TOL_\delta$$

for each $n < n^*$, and hence

$$\max_{n \leq n^*} \| y(t_n) - z_n \| \leq k\, TOL \qquad \text{where } k = c\, \frac{e^{L(t_{n*} - t_0)} - 1}{e^{Lh} - 1},$$

L is the Lipschitz constant of f, and $h = (t_{i+1} - t_i)$.

Since the parallel algorithm consists of approximating the values z_n by iterations u_n^k, the problem is now to give an iteration tolerance TOL_i such that the local error test for iterations leads to a maximum error $\max_{n \le n^*} \| u_n^k - z_n \|$ of the same magnitude of $\max_{n \le n^*} \| y(t_n) - z_n \|$.

In the forthcoming paper [2] it is proved that TOL_δ and TOL_i must be of the same magnitude.

According to the comments following formula (7), the speed up is now expected to be large since the amounts of computation for each individual F_n is possibly very large, expecially for a small tolerance TOL_δ.

In this strategy the two goals, parallelism and accuracy control, have been attained independently, the former by the domain decomposition and the latter by step size varying computation of each F_n.

In the second strategy we propose, the domain decomposition points coincides with the mesh of the one-step ode method. In other words, $F_{n+1}(z_n)$ is the numerical approximation of $y(t_{n+1}, t_n, z_n)$ by a single step of the formula. The target is now to modify, after each iteration, the domain decomposition across which the parallelism takes place, so as to approach a mesh selection suitable for the desired accuracy.

Once again we start from guessed initial values $\{u_0^o = y_o, u_1^o, \dots u_{n^*}^o\}$ on a guessed mesh $M^o = \{t_0^o, t_1^o, \dots t_{n^*}^o\}$. The first stage consists in the parallel computation of $F_{n+1}(u_n^o)$, $n = 0, 1, \dots n^*$, and, via a halfspacing local error estimate according to some given tolerance, in parallel estimates of more appropriate step lengths $h_0^o, h_1^o, \dots h_{n^*}^o$ at the mesh points M^o.

The second stage consists in building up, sequentially, a new mesh $M^1 = \{t_0^1, t_1^1, \dots t_{n^*}^1\}$ by the following inductive rule: assume the mesh points have been computed to $t_k^1 \in [t_s^o, t_{s+1}^o)$, then the next point t_{k+1}^1 is

$$\overset{1}{t}_{k+1} = \overset{1}{t}_k + \frac{\overset{o}{t}_{s+1} - \overset{1}{t}_k}{\overset{o}{t}_{s+1} - \overset{o}{t}_s} \overset{o}{h}_s + \frac{\overset{1}{t}_k - \overset{o}{t}_s}{\overset{o}{t}_{s+1} - \overset{o}{t}_s} \overset{o}{h}_{s+1} .$$

If the rule gives less than n* mesh points, M^1 is filled by further equispaced points.

Afterwards we perform an iteration of the parallel algorithm, across the mesh M^1, by new starting values $\tilde{\mathbf{u}}^o$ obtained from $\mathbf{u}^o$ by interpolation on adjacent points. Unlike the new mesh point,which are obtained via linear interpolation, they must be computed with appropriate accuracy so as to preserve the order of the iteration. For Newton or Steffensen iterations the interpolation must involve at least three points.

The resulting iterate $\mathbf{u}^1$ must be subject to first the local error test for stopping iterations, and second to the halfspacing local error test for accuracy. The segment S_1 of acceptable values is given by the largest segment where both tests succeed. The segment S_2 of steps to be re-iterated has to be selected according to the contraction test (8), and new guess mesh points and respective guess values must be supplied to complete the block of parallelism. Then the next iteration can be executed.

Unlike the first strategy, the concurrent computations are equal through the block steps, and hence the parallel algorithm is suitable for a SIMD machine.

On the other hand some implementative details need to be explored further. In particular, as for the previous strategy, the discretization tolerance TOL_δ must be related to the iteration tolerance TOL_i so as to avoid useless extraiterations. For stiff equations this is a hard problem since, far from the limit (smooth) function the estimator of local error tends to underestimate the step size and then to produce meshes whose steps are much smaller than those which would be appropriate for the limit function.

The trouble seems to be inherent to stiff equations which reveal a fast convergence of the iterates, and this prevents the step selection to adapt to the smoothness of the limit function.

This difficulty is present also in the waveform relaxtion method and has been investigated by Nevanlinna in [10]. Possible ways out to this inconvenience can be found either in the choice of an appropriate ratio between TOL_δ and TOL_i or in overstimating the step size given by the early iteration.

REFERENCES

[1] A.BELLEN, M.ZENNARO. Parallel algorithms for initial value problems for nonlinear difference and differential equations. Quaderni Matematici. Dip. di Sc. Matem. Univ. di Trieste. N. 140 (1987)

[2] A.BELLEN, M.ZENNARO, R.VERMIGLIO. Parallel ODE - solvers with step size control. In preparation.
[3] C.W.GEAR. The potential of parallelism in ordinary differential equations. Department of Computer Sciences U.I.U.C. Rep. # R-86-1246.
[4] C.W.GEAR. Parallel methods for ordinary differential equations. (1987) Preprint.
[5] C.W.GEAR. Massive parallelism across the method in ODEs. Department of Computer Sciences U.I.U.C. Rep # R-88-1442.
[6] P.J.van der HOUWEN, B.P.SOMMEIJER. Variable step iteration of high-order Runge-Kutta methods on Parallel Computers. C.W.I. Amsterdam. Preprint.
[7] P.J.van der HOUWEN, B.P.SOMMEIJER, P.A.van MOURIK. Note on explicit parallel multistep Runge-Kutta methods. C.W.I. Amsterdam. Report NM-R8814.
[8] K.K.JACKSON, S.P.NØRSETT. Parallel Runge-Kutta methods. (1986) A manuscript.
[9] I.LIE. Some aspects of parallel Runge-Kutta methods. Department of Numerical Mathematics. University of Trondheim. 3/87.
[10] O.NEVANLINNA. Remarks on Picard-Lindelöf iteration. Report-MAT-A254. Helsinki University of Technology. Dec. 1987.
[11] U.SCHENDEL. Introduction to Numerical Methods for Parallel Computers. Series in Mathematics and its Applications (1984). John Wiley & Sons.
[12] S.P.NØRSETT, H.H.SIMONSEN. Aspects of parallel Runge-Kutta methods. These Proceedings.
[13] R.D.SKEEL, N.Y.KONG. Blended linear multistep methods. ACM Trans. Math. Software, Dec. 1977, 326-345.

ON THE SPECTRUM OF FAMILIES OF MATRICES WITH APPLICATIONS TO STABILITY PROBLEMS

G.DI LENA D.TRIGIANTE(*)
Dipartimento di Matematica,
Campus Univ., Università di Bari,
70125 Bari, Italy

Abstract. The stability properties of numerical methods for hyperbolic and parabolic PDE are studied by using the method of lines. The notion of spectrum of a family of matrices permits the use of the usual concept of A-stability for ODEs. The problem related to the consistency is analyzed as well as the connection with the Von-Neumann stability test.

1. INTRODUCTION.

The availability of hightly efficient codes for the numerical solution of ordinary differential equations (ODE) has increased the interest in the method of lines for the numerical solution of partial differential equations (PDE). The essence of this method is to discretize all but one of the independent variables and leave the one continuous. One obtains then a system of ODEs which can be solved numerically with an appropriate code. For details, see for example [1], [2], [3].

The method of lines, also called method of Rothe, is, however, far older and it has been used for theoretical purposes (see, for example, [4], [5]).

There is one main difficulty in treating the ODEs arising in the method of lines: the dimension of the system is not fixed, but can become arbitrarily large when the step length of the discretized variables becomes smaller and smaller. As consequence it follows than the usual spectral conditions based on the eigenvalues of the Jacobian may be not enough for the stability of the numerical methods, see [2], [3] and more recently, [5], [6].

It seems to us that the key concept in treating the subject is the notion of the spectrum of families of matrices. This is a restriction of the well known concept of

(*) Work performed within the activities of the "Centro Nazionale di Matematica Computazionale" supported by the "Ministero della Pubblica Istruzione".

the spectrum of operators in Banach spaces.

We will show that the use of this concept makes the study of the stability clearer, reducing in a natural way to the well known Dahlquist theory for ODEs. Moreover, it will also be shown that it contains the popular von Neumann stability test.

2. PRELIMINAIRES.

Let be A an s×s real or complex matrix whose eigenvalues have negative real parts and define y(t) by

$$\frac{dy}{dt} = Ay \quad , \qquad y(0) = y_0. \tag{2.1}$$

With this hypothesis the origin is asymptotically stable. The solution satisfies the relation

$$y(t+\Delta t) = e^{A\Delta t}\, y(t). \tag{2.2}$$

It is known that if f(z) is an analytic function in an open set $\Omega \subset C$, then

$$f(A\Delta t) = \frac{1}{2\pi i} \oint_\Gamma f(z)\,(zI - A\Delta t)^{-1}\, dz \tag{2.3}$$

where Γ is a Jordan curve in Ω that contains in its interior the spectrum of $A\Delta t$. From (2.3) one has

$$e^{A\Delta t} = \frac{1}{2\pi i} \oint_\Gamma e^z\,(zI - A\Delta t)^{-1}\, dz \tag{2.4}$$

Let us consider now a rational approximation R(z) of e^z satisfying the following conditions:

1) $| e^z - R(z) | = 0(z^{p+1}) \qquad p \geq 1$

2) $| R(z) | \leq 1 \qquad z \in D \subset C.$

One has

$$e^{A\Delta t} - R(A\Delta t) = \frac{1}{2\pi i} \oint_\Gamma (e^z - R(z))\,(zI - A\Delta t)^{-1}\, dz.$$

Using $R(A\Delta t)$ we define a one-step method which approximates the solutions of equation (2.1):

$$y_{n+1} = R(A\Delta t)\, y_n.$$

By substracting and letting $e_n = y(t_n) - y_n$ one has:

$$\begin{aligned} e_{n+1} &= e^{A\Delta t}\, y(t_n) - R(A\Delta t)\, y_n = \\ &= R(A\Delta t)\, e_n + (e^{A\Delta t} - R(A\Delta t))\, y(t_n). \end{aligned}$$

On letting

$$(e^{A\Delta t} - R(A\Delta t))\, y(t_n) = b_n,$$

one obtains (we take $e_o = 0$ for simplicity):

$$e_n = \sum_{j=0}^{n-1} R^{n-j-1}\, (A\Delta t)\, b_j.$$

To control the error e_n one must require that $\| R^m\, (A\Delta t) \| \leq k$ for all m and $\| b_j \|$ bounded. Written in integral form one has

$$R(A\Delta t) = \frac{1}{2\pi i} \oint_\Gamma R(z)\, (zI - A\Delta t)^{-1}\, dz.$$

If $\| (zI - A\Delta t)^{-1} \|$ is bounded for $z \in \Gamma$ and Γ is a Jordan curve containing in its interior the eigenvalues of $A\Delta t$ and if on Γ $\quad | R(z) | \leq 1$, that is, Γ is contained in the absolute stability region of the method defined by

$$D = \{ z \in C : | R(z) | \leq 1 \},$$

then

$$\| R(A\Delta t) \| \leq \frac{1}{2\pi}\, s\gamma,$$

and in general for $j \geq 0$

$$\| R^j\, (A\Delta t) \| \leq \frac{1}{2\pi}\, s\gamma$$

where s is the length of Γ and γ a bound for $\| (zI - A\Delta t)^{-1} \|$.

3. SYSTEMS WITH VARIABLE DIMENSIONS.

If the dimension of the system of ODEs is variable, we are faced with the family of problems:

$$\frac{dy_s}{dt} = A_s y_s \quad , \qquad y_s(t_o) = y_s(0),$$

where y_s are vectors defined in R^s (or C^s). As the dimension s increases, we have a family of ODEs defined by the family of matrices $\{A_s\}$. Usually A_s is the restriction to R^s of an operator A defined in a Banach space (see section 4). In this case one needs bounds that are independent of s. As before we have

$$e_n = \sum_{j=0}^{n-1} R^{n-j-1} (A_s\Delta t)\, b_j$$

where

$$R^m(A_s\Delta t) = \frac{1}{2\pi i} \oint_\Gamma R^m(z)\, (zI - A_s\Delta t)^{-1}\, dz$$

$$b_n = \frac{1}{2\pi i} \oint_\Gamma (e^z - R(z))\, (zI - A_s\Delta t)^{-1}\, dz\; y_s(t_n).$$

As in the previous case we need Γ to be a Jordan curve such that

$$\| (zI - A_s\Delta t)^{-1} \| \leq \gamma \qquad \text{for } z \in \Gamma$$

and in this case γ must be independent of s, also. This is not, in general, accomplished by considering only the eigenvalues of the matrices A_s. To obtain the result we must consider the spectrum of the family $\{A_s\}$. Before doing this, we shall introduce in the next section some examples of families of matrices.

4. EXAMPLES OF FAMILIES OF MATRICES.

We shall briefly give three examples of families of matrices by applying the method of lines to three typical problems.

Example 1.

$$\frac{\partial u}{\partial t} = \frac{\partial^2 u}{\partial x^2}$$

$$u(0, t) = u(1, t) = 0$$

$$u(x, 0) = \phi(x).$$

Letting $x_i = i\Delta x$, $i = 0, 1, 2, ..., N+1$, we have

$$\frac{\partial^2 u}{\partial x^2} \simeq \frac{u(x_{i+1},t) - 2u(x_i,t) + u(x_{i-1},t)}{\Delta x^2}$$

$$U(t) = \begin{pmatrix} u(x_1,t) \\ u(x_2,t) \\ \cdot \\ \cdot \\ u(x_N,t) \end{pmatrix}; \quad B_N = \begin{pmatrix} -2 & 1 & 0 & & \\ 1 & -2 & 1 & & \\ & \cdot & \cdot & \cdot & \\ & & \cdot & \cdot & \cdot \\ & & & \cdot & \cdot \end{pmatrix}; \quad \phi = \begin{pmatrix} \phi(x_1) \\ \phi(x_2) \\ \cdot \\ \cdot \\ \phi(x_N) \end{pmatrix},$$

the semidiscretized problem becomes:

$$\frac{dU}{dt} = \frac{1}{\Delta x^2} B_N U \ , \qquad U(0) = \phi.$$

Example 2.

$$\frac{\partial u}{\partial t} = a\frac{\partial u}{\partial x}$$

$$u(0, t) = 0 \ \text{ or } \ u(1, t) = 0, \ \ u(x, 0) = \phi(x).$$

As before the semidiscretized problem is

$$\frac{dU}{dt} = \frac{a}{\Delta x} A_N U \ , \qquad U(0) = \phi,$$

where

$$A_N = \begin{pmatrix} -1 & 1 & 0 & \dots 0 \\ & -1 & 1 & \dots\dots \\ & & \cdot & \cdot \\ & & & \cdot\ \cdot \\ & & & \cdot \end{pmatrix} \quad \text{or } A_N = \begin{pmatrix} 1 & & \\ -1 & 1 & & 0 \\ & \cdot & \cdot & \\ 0 & & \cdot & \cdot \\ & & & \cdot\ \cdot \end{pmatrix}$$

according to the sign of a.

Example 3.

$$\frac{\partial^2 u}{\partial t^2} = \frac{\partial^2 u}{\partial x^2}$$

$$u(0, t) = u(1, t) = 0,$$

$$u(x, 0) = \phi(x)$$

The semidiscretized problem is

$$\frac{d^2U}{dt^2} = \frac{1}{\Delta x^2} B_N U \quad , \quad U(0) = \phi.$$

Letting

$$V = \frac{dU}{dt}, \quad W = \begin{pmatrix} U \\ V \end{pmatrix}, \quad \psi = \begin{pmatrix} \phi \\ 0 \end{pmatrix},$$

the problem can be written as a first order problem:

$$\frac{dW}{dt} = \frac{1}{\Delta x^2} D_{2N} W \quad , \quad W(0) = \psi,$$

where

$$D_{2N} = \begin{pmatrix} 0 & \Delta x^2 I \\ & \\ B_N & 0 \end{pmatrix}_{2N \times 2N}.$$

On noting that

$$D_{2N}^2 = \begin{pmatrix} B_N & 0 \\ & \\ 0 & B_N \end{pmatrix} \Delta x^2,$$

it follows that in all the previous examples one is faced with matrices of the form

$$C_n = \begin{pmatrix} \alpha & \beta & & & 0 \\ \gamma & \alpha & \beta & & \\ & \cdot & \cdot & \cdot & \\ & & \cdot & \cdot & \cdot \\ 0 & & & \gamma & \alpha \end{pmatrix}$$

with $\alpha + \beta + \gamma = 0$. This condition is essentially due to the consistency condition.

5. SPECTRUM OF A FAMILY OF MATRICES.

Let $\{A_n\}$ be a family of complex matrices of increasing dimension n.

Definition 1. The spectrum of the family is the set of complex numbers λ such that for every $\varepsilon > 0$ there exist $n \geq 1$, $x_n \in C^n$, $x_n \neq 0$, such that

$$\| A_n x_n - \lambda x_n \| \leq \varepsilon \| x_n \|.$$

The spectrum of the family $\{A_n\}$ will be denoted by $S(\{A_n\})$ or simply S when no confusion arises. The set of all eigenvalues of all matrices A_n will be denoted by Σ. Of course $\Sigma \subset S$. The elements of $Q = S \backslash \Sigma$ which are not eigenvalues of any matrix A_n will be called quasi eigenvalues.
The set S has important properties.

Proposition 1. $S(\{A_n\})$ is a closed set.

Proof. Let μ_i be a sequence of elements of S coverging to λ. From

$$\| A_n x - \lambda x \| \leq \| A_n x - \mu_i x \| + | \lambda - \mu_i | \| x \|,$$

it follows that fixing the positive number $\varepsilon/2$, there exist $\nu > 0$ such that for $i > \nu$, one has $| \nu - \mu_i | \leq \varepsilon/2$. Moreover, there exist $n > \nu$ and x^* such that

$$\| A_n x^* - \mu_i x^* \| < \frac{\varepsilon}{2} \| x^* \|$$

for which the result follows by taking $x = x^*$. ■

Proposition 2. If the matrices A_n are normal, and the norm used is $\| \cdot \|_2$, then

$$S(\{A_n\}) = \overline{\Sigma(\{A_n\})}.$$

Proof. Let u_n^i $(i = 1, 2, ..., n)$ be the normalized eigenvectors of A_n and μ_i the eigenvalues of the same matrix. For every $\lambda \in S$, we have, letting $x = \Sigma_i a_n^i u_n^i$

$$\| A_n x - \lambda x \| = (\Sigma_i (a_n^i (\mu_i - \lambda))^2)^{1/2} \leq \varepsilon (\Sigma (a_n^i)^2)^{1/2}$$

from which it follows that

$$\min | \mu_i - \lambda | \leq \varepsilon.$$ ■

Proposition 3. Let K be a compact set in the complex plane not containing elements of S. Then

$$\sup_{n; z \in K} \| (zI - A_n)^{-1} \| < \infty.$$

Proof. See Bakhvalov [8]. ■

Proposition 4. If $z \in Q \backslash \Sigma$, then $\sup_n \| (zI - A_n)^{-1} \| = \infty$ and vice versa.

Proof. Consider $z \in Q \backslash \Sigma$. By definition

$$\| (A_n - zI) x_n \| \leq \varepsilon \| x_n \|.$$

By letting $y_n = (A_n - zI) x_n$, we obtain

$$\| y_n \| \leq \varepsilon \| (A_n - zI)^{-1} y_n \| \leq \varepsilon \| (A_n - zI)^{-1} \| \| y_n \|$$

from which

$$\| (A_n - zI)^{-1} \| \geq 1/\varepsilon$$

and

$$\sup_n \| (A_n - zI)^{-1} \| = \infty.$$

Suppose now that

$$\sup_n \| (A_n - zI)^{-1} \| = \infty,$$

there exists a sequence n_k such that

$$\lim_{k\to\infty} \| (A_{n_k} - zI)^{-1} \| = \infty$$

and vectors y_{n_k} with $\| y_{n_k} \| = 1$ such that

$$\| (A_{n_k} - zI)^{-1} y_{n_k} \| = \| (A_{n_k} - zI)^{-1} \|.$$

Letting $x_{n_k} = (A_{n_k} - zI)^{-1} y_{n_k}$ one has

$$\| (A_{n_k} - zI) x_{n_k} \| = 1 = \frac{\| (A_{n_k} - zI)^{-1} y_{n_k} \|}{\| (A_{n_k} - zI)^{-1} \|} = \frac{\| x_{n_k} \|}{\| (A_{n_k} - zI)^{-1} \|}. \quad ■$$

Proposition 5. Let $\{A_n\}$ and $\{B_n\}$ be two families of matrices such that for every $x_n \in C^n$ with $\| x_n \| = 1$ one has

$$\lim_{n\to\infty} \| (A_n - B_n)x_n \| = 0.$$

Then the quasi-eigenvalues of the two families coincide.

Proof. The result follows easily by considering the inequality:

$$\| A_n x - \lambda x \| \leq (A_n - B_n)x \| + \| B_n x - \lambda x \| \leq (\varepsilon_1 + \varepsilon_2) \| x \|. \quad ■$$

Theorem 1. Let $S(\{A_n\})$ be a compact set contained in a simply connected open Ω. If $f(z)$ is an analytic function in Ω, then

$$S(\{f(A_n)\}) = f(S(\{A_n\})).$$

Proof. We show first that for all $x \in C^n$ and for all $\lambda \in S(\{A_n\})$, there exists an $r > 0$ such that

$$\| f(A_n)\, x - f(\lambda)x \| \leq r \, \| A_n\, x - \lambda x \|.$$

Let Γ be a Jordan curve contained in Ω and containing in its interior $S(\{A_n\})$. One has

$$f(A_n) - f(\lambda)I = \frac{1}{2\pi i} \oint_{\Gamma} (f(z) - f(\lambda))\, (zI - A_n)^{-1}\, dz.$$

Since $\dfrac{f(z) - f(\lambda)}{z - \lambda}$ is holomorphic in Ω, one obtains

$$f(A_n) - f(\lambda)I = \frac{1}{2\pi i} \oint_{\Gamma} \frac{f(z) - f(\lambda)}{z - \lambda}\, (z - \lambda)\, (zI - A_s)^{-1}\, dz =$$

$$= \frac{1}{2\pi i} \oint_{\Gamma} \frac{f(z) - f(\lambda)}{z - \lambda}\, (zI - A_n)^{-1}\, dz \quad \frac{1}{2\pi i} \oint_{\Gamma} (z - \lambda)\, (zI - A_n)^{-1}\, dz =$$

$$= \frac{1}{2\pi i} \oint_{\Gamma} \frac{f(z) - f(\lambda)}{z - \lambda}\, (zI - A_n)^{-1}\, dz\, (A_n - \lambda I),$$

from which

$$\| (f(A_n) - f(\lambda)I)\, x \| \leq r \, \| (A_n - \lambda I)\, x \|$$

where

$$r = \sup_n \frac{1}{2\pi} \left\| \oint_{\Gamma} \frac{f(z) - f(\lambda)}{z - \lambda}\, (zI - A_n)^{-1}\, dz \right\|.$$

Since x is arbitrary, we can choose it such that, for $\lambda \in S(\{A_n\})$,

$$\| (f(A_n) - f(\lambda)I)x \| \leq r\, \varepsilon \, \| x \|$$

and then

$$f(S(\{A_n\})) \subseteq S(f(\{A_n\})).$$

Suppose now that

$$S(f(\{A_n\})) \not\subseteq f(S(\{A_n\})) \tag{4.1}$$

There exists a $\mu \in S(f(\{A_n\}))$ such that $\mu \notin f(S(\{A_n\}))$. Let $B(\mu, r)$ be a circle of center μ and radius r such that $B(\mu, r) \cap f(S(\{A_n\})) = \varnothing$, and let Γ_1 be the regular boundary of a domain containing $S(\{A_n\})$ in its interior. The operator

$$G(A_n) = \frac{1}{2\pi i} \oint_{\Gamma_1} \frac{1}{f(z) - \mu} (zI - A_n)^{-1} \, dz$$

is well defined for each value of n, since Γ_1 does not contain elements of $S(\{A_n\})$ and the function $(f(z) - \mu)^{-1}$ is analytic in the domain. One has then

$$G(A_n) = (f(A_n) - \mu I)^{-1}.$$

On the other hand one has $\mu \in S(f(\{A_n\}))$ and for this reason $G(A_n)$ should be unbounded with respect to n. A contradiction follows so such a μ cannot exist. ■

Definition 2. The quantity

$$\rho = \sup_{\lambda \in S} | \lambda |$$

will be called the spectral radius of the family.

Proposition 6. If ρ is bounded, then for $r > \rho$, there exists a $k > 0$ such that

$$\| A_n^m \| \leq r^{m+1} k.$$

Proof. Let Γ be the set $| z | = r$. Then

$$\| A_n^m \| = \frac{1}{2\pi} \| \oint_{\Gamma} z^m (zI - A_n)^{-1} \, dz \| \leq r^{m+1} \sup_{n; z \in \Gamma} \| (zI - A_n)^{-1} \|$$

and, by Proposition 3,

$$\| A_n^m \| \leq r^{m+1} k.$$ ■

Corollary. If $\rho < 1$, the powers A_n^m are bounded in norm, uniformly with respect to n and m.

Proposition 7. If $\rho > 1$, then $\sup \| A_n^m \| = \infty$.

Proof. Let $z \in S$ and $| z | > 1$. If z is an eigenvalue of A_n, it follows that

$$\| A_n^m \| \to \infty \qquad \text{for } m \to \infty.$$

If $z \in Q$, then for $\varepsilon = 1$ one has by definition

$$\| (A_n - zI)x_n \| \le \| x_n \|,$$

from which $| z | \le 1 + \| A_n \|$. Since from Theorem 1, $z^m \in S(\{A_n^m\})$, there exists an n such that

$$| z^m | \le 1 + \| A_n^m \|$$

and by $| z | > 1$ it follows that

$$\sup \| A_n^m \| = \infty.$$

■

It can happen that the spectral radii of the matrices A_n are less than one for all n and $\rho > 1$. In this case the norms of the powers are not bounded uniformly with respect to both m and n.

6. USE OF THE SPECTRUM.

Once the spectrum of a family has been obtained the conditions of stability can be established. Let

$$D = \{z \in C : | R(z) | \le 1\}$$

and suppose that D contains the spectrum S of the family $\{A_n\}$. Moreover, let Γ be a Jordan curve contained in D and containing S in its interior. One has

$$R^m(A_n\Delta t) = \frac{1}{2\pi i} \oint_\Gamma R^m(z)\,(zI - A_n\Delta t)^{-1}\,dz.$$

Since, from proposition 3

$$\| (zI - A_n\Delta t)^{-1} \| < \gamma,$$

we obtain

$$\| R^m(A_n\Delta t) \| \le \frac{\gamma}{2\pi}\, s$$

s being the length of Γ; that is, the powers $R^m(A_n\Delta t)$ are bounded in norm, uniformly in m and n.

Unfortunately, the consistency conditions on the discretizations imply that the two sets S and the boundary of D have a point (the origin) in common. This implies that any Jordan curve Γ must have the same point in common with the two sets S and the boundary of D and in general the previous result becomes only a necessary condition. Let us give a simple example. Consider the following family of matrices:

$$A_n\Delta t = q \begin{pmatrix} -1 & 1 & & & \\ & -1 & 1 & & \\ & & \cdot & \cdot & \\ & & & \cdot & \cdot \\ & & & & \cdot \end{pmatrix}_{n\times n}$$

with $q > 0$. One has

$$(zI - A_n\Delta t)^{-1} = \begin{pmatrix} \frac{1}{z+q} & \frac{q}{(z+q)^2} & \cdots & \cdots & \frac{q^{s-1}}{(z+q)^s} \\ & \cdot & \cdot & \cdot & \cdot \\ & & \cdot & \cdot & \cdot \\ & & & \cdot & \cdot \\ & & & & \frac{1}{z+q} \end{pmatrix}$$

Let $R(z) = (1 + z)$. The set D is the circle with center in (-1, 0) and radius 1. We must then impose $q \leq 1$ in order to have $S \subseteq D$. *For all values of q the origin is common to the two sets S and the boundary of D.* For $z = 0$ the norm of $(zI - A_n\Delta t)^{-1}$ is given by n/q, that is, it is dependent on the dimension n. Note that the integral

$$\oint_{|z+1|=1} (zI - A_n\Delta t)^{-1}\, dz = I$$

has a norm independent of n, even though for $z = 0$ the norm of the integrand does depend on n.

The question arises:

For which functions $R(z)$ *and for which matrices* A_n *is the norm of the integral*

$$\oint_{\Gamma} R^m(z)\, (zI - A_n)^{-1}\, dz$$

independent of n *and* m?

It is difficult to answer this in general. There are, however, cases that can be answered. For example if the matrices are normal, then the norms of $(zI - A_n)^{-1}$ are bounded. In fact one has:

Theorem. If the matrices A_n are normal, and $S \subseteq D$, then

$$\| R^m(A_n\Delta t) \| \leq 1.$$

Proof. In this case $\| R(A_n\Delta t) \| \leq \max_{z\in S} | R(z) | \leq 1.$ ■

The condition $S \subseteq D$ is in this case both necessary and sufficient for the stability.

One can get to the same result for this example by doing the calculation directly. In fact let f(z) be an analytic function in S such that $f(z) = \sum a_i(z+q)^i$. One has

$$f(z)\,(zI - A_s\Delta t)^{-1} = \begin{pmatrix} \sum_{i=-1} a_{i+1}(z+q)^i & q\sum_{i=-2} a_{i+2}(z+q)^i & \cdots \\ & \cdot & \cdot \\ & & \cdot \end{pmatrix}.$$

One obtains:

$$\oint_\Gamma f(z)\,(zI - A_n\Delta t)^{-1}\,dz = \begin{pmatrix} a_0 & a_1 q & \cdots & a_{s-1}q^{s-1} \\ & \cdot & \cdots & \\ & & \cdot & \cdot \\ & & & a_0 \end{pmatrix}$$

and the norm

$$\| \oint_\Gamma f(z)\,(zI - A_n\Delta t)^{-1}\,dz \|_\infty = \sum_{i=0}^{n} | a_i | q^i.$$

Now let $f(z) = (1+z)^m = \sum_{j=0}^{m} \binom{m}{j} (1-q)^{m-j} (z+q)^j$. One has, for $q \leq 1$,

$$\sum_{i=0}^{m} | a_i | q^i = \sum_j \binom{m}{j} (1-q)^{m-j} q^j = 1.$$

Then the norm is uniformly bounded for $q \leq 1$.
A similar result hold true by using $\| \cdot \|_1$ and then $\| \cdot \|_2$.

A much shorter proof can be given by considering that in the present case the condition $q \leq 1$ implies that

$$\| I + q A_n \|_2 = [\rho((I+qA_n^T)(I+qA_n))]^{1/2} \leq 1,$$

and of course this implies the stability.

The last result obtained for the matrices $I+qA_n$ can be obtained as a special case of a more general lemma due to von Neumann which we shall state in a slightly modified form.

Lemma (von Neumann). Let A be an operator in an Hilbert space. The following conditions are equivalent:

1) $\| (I + A)(I - A)^{-1} \| \leq 1$
2) $\mathrm{Re}\,(Ax, x) \leq 0$
3) $\| R^m(A) \| \leq 1$ for all m if $| R(z) | \leq 1$ when $\mathrm{Re}\,(z) \leq 0$.

The proof can be found in [10].

Suppose now that the spectrum of A is contained in the circle $B(-\rho, \rho)$ with center at $(-\rho, 0)$ and radius ρ, excluding the point $(-2\rho, 0)$. The function

$$w = p(z) \equiv \frac{z}{z+2\rho},$$

transforms the circle into the left half complex plane. It follows that the spectrum of

$$p(A) = (A + 2\rho I)^{-1} A$$

is contained in the negative half plane.

Theorem. If R(z) is a rational function such that $| R(z) | \leq 1$ for $z \in B(-\rho, \rho)$, then the following conditions are equivalent:

1) $\| A + \rho I \| \leq \rho$
2) $A + A^T + \rho^{-1} A^T A$ is negative definite
3) $\| R^m(A) \| \leq 1$ for all m.

Proof. The inverse transform p^{-1} is well defined and

$$\| R^m(A) \| = \| R^m(p^{-1}(p(A))) \|.$$

The operator p(A) has its spectrum in the left half plane and $R(p^{-1}(z))$ is a rational function such that

$$| R(p^{-1}(z)) | \le 1 \text{ for } \mathrm{Re}\,(z) \le 0.$$

The lemma can be applied with p(A) instead of A and $R(p^{-1}(z))$ instead of R(z), to obtain

1) $\| (I+p(A))\,(I-p(A))^{-1} \| \le 1$
2) $\mathrm{Re}\,(p(A)x, x) \le 0$
3) $\| R^m(A) \| \equiv \| R^m(p^{-1}(p(A)) \| \le 1$

which are equivalent to the ones given in the theorem. For $\rho \to \infty$ one obtains the result used in the literature (see [6], [7]). ■

7. SPECTRUM AND VON NEUMANN STABILITY TEST.

The most used test for stability, the so-called von Neumann test, can be shown to be equivalent to asking that the set $S \setminus \Omega$ of a family be contained in the unit circle of the complex plane. In fact, let

$$y_{n+1} = A_N\, y_n, \qquad\qquad A_N \in C^{N \times N}$$

be the discretization of a continuous problem. The von Neumann test amounts to considering solutions of the type

$$y_n = \begin{pmatrix} \lambda^0 \\ \lambda^1 \\ \cdot \\ \cdot \\ \lambda^{N-1} \end{pmatrix} \cdot \xi^n \equiv \Lambda_N\, \xi^n$$

with $| \lambda | = 1$ and $| \xi | < 1$. By substituting one has

$$A_N\ \Lambda_N\ - \xi\, \Lambda_N = 0.$$

The set defined by the values of ξ allowed is a subset of the quasi eigenvalues. In fact, written row by row, the previous equation defines the same difference equation used to obtain the quasi eigenvalues.

8. DETERMINATION OF THE SPECTRUM OF $\{C_N\}$.

Suppose for simplicity that $|\gamma/\beta| < 1$. The following theorem can be proved (see [2] or [3]).

Theorem. The set of the complex numbers λ such that the zero solution of the difference equation

$$\beta x_{i+1} + (\alpha - \lambda) x_i + \lambda x_{i-1} = 0$$

$$x_1 = 1, \; x_2 = \frac{\lambda - \alpha}{\beta}$$

is asymptotically stable is contained in $S(\{C_n\})$.

In the present case this set is essentially the set S. To get Σ we must consider the eigenvalues, that is, the values of λ which solve the boundary value problem obtained from the previous initial value problem by imposing $x_{n+1} = 0$ for all n. In the more general case the set obtained in the theorem is a subset of S to which one must add the set Σ. Particular cases are:

1) $\alpha = -2, \beta = \gamma = 1$. The spectrum is the segment $-2 + 2\cos\theta$ with $0 \le \theta \le \pi$.
2) $\alpha = -1, \beta = 1, \gamma = 0$. The spectrum is the circle $x = 1 - \cos\theta$, $y = \sin\theta$.

REFERENCES

[1] L.F.SINCOVEC, N.K.MADSEN. Software for nonlinear partial differential equations. ACM Trans. Math. Soft. 1(1975) 231-263.
[2] G.DI LENA, D.TRIGIANTE. On the stability and convergence of the lines method. Rendiconti di Matem. 3(1983) 113-126.
[3] G.DI LENA, D. TRIGIANTE. Il metodo delle linee in Analisi Numerica. Rend. Sem. Mat. Univ. Pol. Torino, 42(1984) 25-41.
[4] O.A.LISKOVETS. The Method of Lines (Review). Diff. Eq. 1(1965) 1308-1323.

[5] W.WALTER. The line method for parabolic differential equations. Problems in boundary layer theory and existence of periodic solutions. Lecture Notes in Math. N. 430, Springer-Verlag 1974.

[6] M.N.SPIJKER. Stepsize restriction for stability of one-step methods in the numerical solution of initial value problems. Math. of Comp. 45(1985) 377-392.

[7] J.M.SANZ-SERNA, J.G.VERWER. Convergence analysis of one-step schemes in the method of lines. Report NM-R8608, Centre for Math. and Comp. Science, Amsterdam (1986).

[8] N.S.BAKHVALOV. Numerical Methods. MIR, Moscow (1977).

[9] S.K.GODUNOV, V.S. RYABENKI. Theory of Difference Schemes. North-Holland (1964).

[10] F.RIESZ, B.SZ-NAGY. Leçons d'Analyse functionnelle. Budapest (1953).

DAEs: ODEs WITH CONSTRAINTS AND INVARIANTS

C.W.GEAR (*)
Department of Computer Science
University of Illinois at Urbana-Champaign
U.S.A.

Abstract. This paper is an extension of parts of two earlier papers [3, 4] which dealt with the problems of maintaining invariants in the numerical solution of ODEs and the index reduction of differential-algebraic equations (DAEs). DAEs with their constraints ignored can be viewed as undeterminated ODEs, while ODEs with their invariants appended can be viewed as overdetermined DAEs. Index reduction applied to DAEs appends additional equations derived by differentiating some of the equations to get an overdetermined system. Methods for maintaining invariants in the solution of ODEs are extended to DAEs to handle this case.

1. INTRODUCTION.

The general differential-algebraic equations is written as

$$F(y', y, t) = 0 \tag{1}$$

where $y(t) \in \mathbf{R} \rightarrow \mathbf{R}^n$, and $F \in \mathbf{R}^n \times \mathbf{R}^n \times \mathbf{R} \rightarrow \mathbf{R}^n$ is a function for which we will assume sufficient differentiability. (März [8] considers systems in which discontinuities are permitted). An important special form of differential-algebraic equations is the *semi-explicit* form

$$u' = f(u, v, t) \tag{2}$$

$$0 = g(u, v, t) \tag{3}$$

(*) Work supported in part by the US Department of Energy under grant DOE DEFG02-87ER25026.

considered by Petzold and Lotstedt [7, 9] and Brenan [1]. We will give an equivalence between semi-explicit and general differential-algebraic equations so that we can discuss the nature of the latter in order to understand the structure of all DAEs.

In the semi-explicit form, the first part, eq. (2), looks like a differential equation with driving functions, v, while the second part, eq. (3), looks like *constraints* on the driving functions. Normally u and f are in $\mathbf{R}^{n1}$ and v and g are in $\mathbf{R}^{n2}$ so there are just enough constraints to determine the driving functions if appropriate conditions are satisfied. Problems in which the number of constraints exceeds the number of driving functions are of interest. In this case, consistency conditions must be satisfied, so that some of the constraints could be discarded because they are invariants of the solution of the remaining equations. However, it may not be practical to discard these invariants because the invariants may be integral relations that would not be satisfied by a numerical solution in the presence of roundoff and truncation errors. As an example of such a situation, consider the ODE

$$y' = f(y, t). \tag{4}$$

If a function $h(y, t)$ is such that

$$h_y f + h_t \equiv 0 \tag{5}$$

then h is an integral invariant along any solution of eq. (4) because

$$h(y, t) = h(y(t_0),t_0) + \int_{t_0}^{t} (h_y f + h_t)\, dt = k \tag{6}$$

where k is a constant. However, a numerical method will not, in general preserve k. Therefore, we are interested in method for solving overspecified differential-algebraic equations directly. We will present a method that introduces additional variables to remove the overdeterminacy, but which is such that all solutions of the enlarged system are solutions of the original system and such that a numerical method applied to the enlarged system will produce solutions that preserve the invariants.

The index, m, of the DAE (1) is defined as follows:

m = 0: If $\partial F/\partial y'$ is non-singular, the index is 0. Under this condition, eq. (1) can, in principle, be inverted into the explicit ODE $y' = f(y, t)$ so in this case we call eq. (1) an implicit ODE.

m > 0: Consider the systems of equations

$$F(y', y, t) = 0$$

$$\frac{dF}{dt} = \frac{\partial F}{\partial y'} y^{(2)} + \frac{\partial F}{\partial y} y' + \frac{\partial F}{\partial t} = 0$$

$$\frac{d^2F}{dt^2} = \frac{\partial F}{\partial y'} y^{(3)} + \ldots = 0 \tag{7}$$

$$\ldots$$

$$\frac{d^sF}{dt^s} = \frac{\partial F}{\partial y'} y^{(s+1)} + \ldots = 0$$

as a system of equations in the *separate* dependent variables y', $y^{(2)}$, ..., $y^{(s+1)}$, and solve for these variables as functions of y and t considered as *independent* variables. Since $\partial F/\partial y'$ is singular, it will not be possible to solve for $y^{(s+1)}$, or possibly even for other y^q with smaller q. If it is possible to solve for y' for some finite s, then the index, m, is defined as the smallest integer s for which eqs (7) can be solved for

$$y' = y'(y, t) \tag{8}$$

This equation is called the *underlying* ODE.

The numerical solution of high-index problems is difficult. The definition of index indicates a construction by which one can obtain a differential equation. Any solution of the original DAE (1) is a solution of its underlying ODE. However, solutions of the underlying ODE are not necessarily solutions of the original DAE because the differentiation has introduced constants of integration. A simple example of this is the index one DAE $y - t^2 = 0$. Its underlying ODE is $y' = 2t$ whose solutions are $y = t^2 + c$ for any constant c. However, the pair of equations ($y - t^2 = 0$, $y' = 2t$) can be viewed as an overspecified DAE which is consistent and has just the solutions of the original problem. We will examine ways to reduce the index of a DAE by differentiation, obtaining an overspecified DAE, and then use the techniques for overspecified DAEs to produce a lower-index DAE than the original one that has the same set of soutions as the original one and no additional solutions.

2. EQUIVALENCE OF GENERAL AND SEMI-EXPLICIT SYSTEMS.

We will give simple tranformations between the general and semi-explicit forms. The general form (1) is equivalent to a semi-explicit DAE of index one higher (which suggests that the index definition for semi-explicit forms may not be the most sensible), and vice versa.

We transform the general form into the semi-explicit form by replacing y' with v, then y with u to get

$$u' = v \tag{9}$$

$$0 = F(v, u, t) \tag{10}$$

which is a semi-explicit system. Its index is no more than one greater than the index, m, of (1) because, by differentiation of (1) m times we can solve for $v = u'(u, t)$, and with one additional differentiation we can compute $v' = v'(u, t) = \partial u'/\partial t + +(\partial u'/\partial u)u'(u, t)$.

We transform the semi-explicit form (2) - (3) into a general form by replacing u by y_1 and v by y'_2 to get

$$y'_1 - f(y_1, y'_2, t) = 0 \tag{11}$$

$$g(y_1, y'_2, t) = 0 \tag{12}$$

If the index, m, of (2) - (3) is zero, the dimension, n_2, of v and g is zero and the index, m', of (11) is also zero. Otherwise, in m' differentiations of (11) - (12) we can form $y'_2(y_1, t)$. Thus, in m' differentiations of (2) - (3) we can form v(u, t), and one more will give us v'(u, t). Thus, $m \le m' + 1$. Conversely, we can compute v'(u, v, t) by m differentiations of (2) - (3), so in the same number of differentiations of (11) - (12) we can get $y_2^{(2)}(y_1, y'_2, t)$. Thus, $m' \le m$. Usually, we can get y'_2 in one less differentiation: in that case, $m' = m - 1$.

We conjecture that the conversion between from the general to the semi-explicit form, and the reverse, always raises or lowers the index by one, respectively, except in the special case of a semi- explicit form that is an explicit ODE, *in which case there is no change of index.*

Although the transformation from (2) - (3) to (11) - (12) has reduced the index, we still have to differentiate y_2 if we want to recover the original variable v. Hence, the number of differentiations that are effectively present has not changed (and it is this number which can have a significant impact on the numerical calculations).

In the transformation from the general to the semi-explicit form, the increase in the index is due to the introduction of additional variables. Suppose we start with a semi-explicit equation and simply view it as a general form, that is, as

$$y'_1 - f(y_1, y_2, t) = 0 \tag{13}$$

$$g(y_1, y_2, t) = 0 \tag{14}$$

which has the form F(y', y, t), and then formally transform it to semi-explicit form. We get

$$u'_1 = v_1$$
$$u'_2 = v_2$$
$$v_1 - f(u_1, u_2, t) = 0$$
$$g(u_1, u_2, t) = 0$$

This will formally increase the index by one, but only because we have introduced additional variables v_1 and v_2. The definition of index requires the differentiation of these variables, although they are not variables in the original problem, but correspond to $u^{(2)}$ and $v^{(2)}$. If we avoid computing these values in a numerical procedure, there is no reason for it to encounter difficulties due to a higher index.

We see that the semi-explicit form can be transformed to the general form either in eqs (11) - (12) or eqs (13) - (14). If we had a general form that had the structure of eqs (13) - (14), we could transform it to the general form of lower index in eqs (11) - (12). This leads us to the definition of the *minimum index equivalent* of (1) as follows: For each j, if the j-th column of $\partial F/\partial y'$ is identically zero replace the variable y_j with the variable y'_j. The index of the resulting system will be called the *minimum index* of the system. The minimum index of a system is invariant under any of the transformations described above.

3. MAINTAINING INVARIANTS.

Many systems of ordinary differential equations that are solved numerically have known invariants, that is, relations that hold along any solution trajectory. These are frequently quantities such as the total energy, momentum, mass, etc., of the systems, and as such are partial integrals of the system. Numerical methods do not always maintain the constancy of these invariants, but it is often desiderable to conserve them because the solution of the problem might be quite sensitive to small changes in their value and might even become unstable if they are permitted to change. We refer the reader to Shampine [10] for a discussion of these reasons and a review of other literature. In this section we consider ways in which invariants of both ODEs and DAEs can be enforced during the numerical computation of their solution.

Invariants mainly occur in ODEs, but we will find them to be important in the reduction of the index of DAEs by differentiation. Important examples of invariants in ODEs include both linear and nonlinear invariants. As examples of linear invariants we consider the D3 system of equations from the stiff test set [2]. This system is

$$y'_1 = y_3 - 100\, y_1 y_2 \tag{15}$$

$$y'_2 = y_3 + 2y_4 - 100\, y_1 y_2 - 2 \times 10^4\, y_2^2 \tag{16}$$

$$y'_3 = -y_3 + 100\, y_1 y_2 \tag{17}$$

$$y'_4 = -y_4 + 10^4\, y_2^2 \tag{18}$$

These equations describe a chemical kinetic problem in which y_1 and y_2 react to form y_3, and y_2 reacts with itself to form y_4. It can be seen that the mass-balance invariants ot this system are

$$k_1 = y_1 + y_3 \tag{19}$$

and

$$k_2 = y_2 + y_3 + 2y_4 \tag{20}$$

Energy is a common example of a nonlinear invariant. Consider a simple pendulum described by

$$\theta' = \omega \tag{21}$$

$$\omega' = -g \sin(\theta)/L \tag{22}$$

where θ is the angle with the vertical, L the pendolum length, and g the acceleration due to gravity. Eqs (21) - (22) have a constant energy, so we have the invariant

$$k = \omega^2 - 2g \cos(\theta)/L \tag{23}$$

Any invariant of an ODE of the form $h(y, t) = k$ is an *integral invariant,* that is, it is an integral of some identity derived directly from the differential equation. In DAEs, we restrict ourselves to integral invariants. Since we can always write a DAE in semi-explicit form, we will start with eqs (2) - (3) plus then n_3 additional integral invariants

$$h(u, v, t) = k \tag{24}$$

If the vectors v and g are null (zero dimension), we have the case of an ODE. For h to be an integral invariant of the DAE, we must have that

$$h_u f + h_v v' + h_t \equiv 0 \tag{25}$$

is implied by (3) and its first derivative, that is to say, there must exist a matrix B such that

$$Bg_v + h_v = 0 \tag{26}$$

and the identity

$$[h_u + Bg_u]\, f + h_t + Bg_t \equiv 0 \tag{27}$$

is implied by (3). (Relations (26) and (27) coupled with (3) imply (25) which can then be integrated to yield $h = k$). Note that if $h(u, v, t) = k$ is a direct consequence of (3) then $h_u + Bg_u \equiv 0$ and $h_t + Bg_t \equiv 0$ and the left-hand side of (27) is zero independently of f. In this case, h is simply dependent on g, so any method that satisfies $g = 0$ will automatically satisfy $h = k$. In this case, h is not an integral invariant; it is simply a restatement of some of the constraints (3). DAEs of index greater than one also have differential invariants, that is, ones obtained by differentiation of the system. For example, the index 2 DAE

$$x = g(t)$$

$$x' = y$$

has the differential invariant $y - g'(t) = 0$. However, differential invariants are automatically satisfied in a convergent numerical integration within the truncation and roundoff errors of numerical differentiation. Hence, we will only consider integral invariants.

Define the matrix L to be $h_u + Bg_u$. We change eq. (2) to be

$$u' = f(u, v, t) + L^T \mu \tag{28}$$

and consider the solution of the $n_1 + n_2 + n_3$ equations (28), (3) and (24). We call these the *extended equations*. Numerically, the invariant (24) is guaranteed to be satisfied because it is part of the system to be solved.

We now show that $L^T \mu$ is identically zero in the exact solution of the extended equations, so that these have the same solution as the original DAE (2) - (3). This follows by differentiating (24), eliminating v' using (3) and substituting (28) to get

$$LL^T \mu + Lf + [h_t + Bg_t] = 0. \tag{29}$$

Substituting eq. (27) which is implied by (3) we get

$$LL^T \mu = 0.$$

Premultiplying by μ^T we see that $L^T \mu = 0$ as claimed.

In most cases, the invariant h is independent of v so that $L = h_u$ and eq. (27) is simply $L^T \mu + h_t \equiv 0$.

4. SOLUTION OF THE EXTENDED EQUATIONS.

The reader may wonder if we have not made a simple problem more complex by increasing the number of variables and extending the system of equations. Fortunately, the extended system frequently can be solved with little effort because they can be solved with many of the same techniques used to solve the original DAE or ODE in the absence of the invariants with one additional step to satisfy the invariants. We consider methods for ODEs or DAEs in a semi-explicit form which compute an approximation to the derivate and solution value, u'_n and u_n, at the new time t_n using an implicit relation of the form

$$u_n = \Sigma_n + h\beta u'_n. \tag{30}$$

Multistep methods and many others can be written in this form. These are used with the original equations (2) and (3) to compute an approximation at t_n, say $u_{n,o}$ and $v_{n,o}$ which does not necessarily satisfy the invariant h. Next, the additional term $L^T\mu$ is added to u'_n which produces a correction to (30) given by

$$u_n = u_{n,o} + h\beta L^T\mu \tag{31}$$

where μ is chosen such that

$$h(u_n, v_n, t_n) = 0 \tag{32}$$

Unfortunately, this could require a simultaneous change in v to maintain the identity (3). In the case of ODEs, this is not necessary since g is null. Neither is it necessary in those DAEs for which $g_u L^T = 0$. For these problems (31) - (32) is the only additional step needed.

5. INDEX REDUCTION.

Differentiation reduces the index of a problem, so it is possible, in principle, to reduce the index of a system to zero (or just to a lower index system that can be more easily solved numerically). However, each differentiation introduces additional constants of integration, and, even if these constants can be computed, the numerical integration method may not keep them constant. The index of the semi-explicit system (2) - (3) can be reduced by one by differentiating (3) to get

$$g_u f(u, v, t) + g_v v' + g_t = 0 \tag{33}$$

and using this in place of (3). To remove the constants of integration we can consider (2), (3) and (33) simultaneously, but we then have an overdetermined system although it is consistent. We will use the technique for invariant enforcement to maintain the original "algebraic" conditions in (3) on the differentiated system (2) and (33).

For generality, we start with eq. (1). The procedure below formally differentiates (1) to get an index zero system, and then re-introduces the algebraic constraints. We emphasize that this procedure may not always be practical, but that in many problems the structure of the equations is such that the manipulation is simple.

Let r_1 be the rank of $\partial F/\partial y'$ in (1). If it is n, the dimension of the system, the index is already zero and there is nothing to do. Otherwise, there exists a non-singular $r_1 \times r_1$ submatrix of $F_{y'}$ which is non-singular. Let us suppose that rows and variables have been numbered so that this submatrix is the upper principal minor. Then, by the implicit function theorem, the first r_1 equations in (1) can be solved for the first r_1 components of y'. Let us break y into $(y_1^T, v_1^T)^T$, with the dimension of $y_1 = r_1$, so that y_1 is given by

$$y'_1 = e_1(y_1, v_1, v'_1, t) \tag{34}$$

We can substitute this into the last n - r_1 equations in (1) to get an implicit relationship betweeen y_1, v_1 and t. It cannot involve v'_1 or we would be able to solve (1) for additional components of y'_1, contrary to the assumption about the rank of $F_{y'}$. Thus, the remaining equations can be written as

$$g_1(y_1, v_1, t) = 0 \tag{35}$$

If we differentiate (35) we get

$$g_{1,y_1} e_1(y_1, v_1, v'_1, t) + g_{1,v_1} v'_1 + g_1, t = 0 \tag{36}$$

Equations (34) and (35) are equivalent to the original system (1), while equations (34) and (36) are of one index lower.

We now proceed with the new system and reduce the index further unless it is already zero. Let r_2 be the rank of $A_1 = g_{1,y_1} e_{v'_1} + g_{1,v_1}$. As before, we can find an r_2-squared non-singular minor of A_1 and we will assume that the numbering is such that it is the leftmost principal minor. We break v_1 into $(y_2^T, v_2^T)^T$, where the dimension of y_2 is r_2. Then, by the implicit function theorem, we can solve (36) for y'_2 to get

$$y'_2 = e_2(y_1, y_2, v_2, v'_2, t) \tag{37}$$

Note that this can be used ro eliminate y'_2 from (34). If $r_2 = n - r_1$, the index is now zero and there is nothing more to do. Otherwise, we can substitute this into the last $n - r_1 - r_2$ equations of (36) to get a relation between y_1, y_2, v_2 and t:

$$g_2(y_1, y_2, v_2, t) = 0 \tag{38}$$

Again, this cannot depend on v'_2 because of the assumption on r_2. We can continue this process, each time differentiating the algebraic equation of the form (38), solving for the derivates of additional y'_i variables, and eliminating their derivatives from the earlier equations of the form (34) and (37). If the index is m, the process will terminate after m steps. At this point we will have the set of underlying differential equations of the form

$$y'_i = f_i(y, t) \tag{39}$$

or, in other words, $y' = f(y, t)$. The solutions of these equations include the solutions of the original system (1), but they include additional solutions because of the constants of integration. We need to append to these the algebraic relations (35), (38), etc., that were developed along the way, namely

$$g_1(y_1, v_1, t) = 0 \tag{40}$$

$$g_2(y_1, y_2, v_2, t) = 0 \tag{41}$$

$$\ldots \tag{42}$$

$$g_m(y_1, y_2, \ldots, y_m, v_m, t) = 0. \tag{43}$$

Recognizing that each of these has y as an argument, we can write them simply as

$$g(y, t) = 0 \tag{44}$$

where $g = g_i$. We want to append these equations to (39), but that will make (39) overdetermined. Therefore, we introduce an additional vector variable $\mu = \{\mu_i\}$ of the same dimension as g, and add the term $g_{y_i}^T \mu$ to the right-hand side of (39) to get

$$y'_i = f_i(y, t) + g_{y_i}^T \mu \tag{45}$$

The process described above has reduced the problem to the semi-explicit form (45) and (44) where the "algebraic" variable μ appears linearly. Furthermore, the index of the new system is two since, with one differentiation we can determine that $\mu = 0$, and hence can determine μ' in two differentiations. We could reduce the index to one by transforming to the minimal form, that is, by replacing μ by v' to get

$$y' = f(y, t) + G^T v' \tag{46}$$

$$g(y, t) = 0 \tag{47}$$

where $G = g_y$. However, this minimal form has no apparent computational advantage over the index-2 form.

The reduction process could be stopped after only q differentiations to get the system

$$y'_i = \tilde{e}_i(y_1, \ldots, y_{q+1}, v_{q+1}, v'_{q+1}, t), \qquad i = 1, \ldots, q+1 \tag{48}$$

$$g_{q+1}(y_1, \ldots, y_{q+1}, v_{q+1}, t) = 0 \tag{49}$$

which can be extended to

$$y'_i = \tilde{e}_i(y_1, \ldots, y_{q+1}, v_{q+1}, v'_{q+1}, t) + \sum_{j=1}^{q} \frac{\partial g_j}{\partial y_i} \mu_j \tag{50}$$

$$g_i(y_1, \ldots, y_i, v_i, t) = 0, \qquad i = 1, \ldots, q+1. \tag{51}$$

These are an index max(m - q, 2) DAE whose solution set is the solution set of the original problem.

6. EXAMPLES.

It is instructive to examine two examples. The first is the linear, time-dependent index 2 DAE given in Gear and Petzold [6] to show that the backward Euler method may not work for index 2 problems. It is

$$\begin{pmatrix} 1 & \eta t \\ 0 & 0 \end{pmatrix} y' + \begin{pmatrix} 0 & 1+\eta \\ 1 & \eta t \end{pmatrix} y = \begin{pmatrix} p_1 \\ p_2 \end{pmatrix} \tag{52}$$

or

$$y'_1 = -(1+\eta)y_2 - \eta t y'_2 + p_1 \tag{53}$$

$$y_1 + \eta t y_2 - p_2 = 0 \tag{54}$$

This is already in the form of (34) and (35) so we differentiate (54) and substitute from (53) to get

$$y_2 - p_1 + p'_2 = 0 \tag{55}$$

which is still an algebraic equation. We differentiate again and substitute into (53) to get the system of ODEs

$$y'_1 = -(1+\eta)y_2 - \eta t(p'_1 - p_2^{(2)}) + p_1 \tag{56}$$

$$y'_2 = p'_1 - p_2^{(2)} \tag{57}$$

We couple the algebraic equations (54) and (55) to these introducing the μ variables to get

$$y'_1 = -(1+\eta)y_2 - \eta t(p'_1 - p_2^{(2)}) + p_1 + \mu_1 \tag{58}$$

$$y'_2 = p'_1 - p_2^{(2)} + \eta t\, \mu_1 + \mu_2 \tag{59}$$

The system of four equations (58), (59), (54) and (55) has the solution $\mu = 0$ and y corresponding to the solution of the original problem. It is a semi-explicit, index two problem which can be solved numerically.

Our second example consists of a set of Euler-Lagrange with constraints. Such a system always has index three and can easily be written in semi-explicit from. The particular case we discuss is a model of a simple pendulum. The equations, after normalization, are

$$x' = u \tag{60}$$

$$y' = v \tag{61}$$

$$u' = -Tx \tag{62}$$

$$v' = -Ty - 1 \tag{63}$$

$$x^2 + y^2 = 1 \tag{64}$$

If we apply the process above, we first differentiate (64) and substitute for derivates from the earlier equations. This process is repeated three times until a differential equation for T is obtained. In the process we get two additional algebraic equations, namely

$$xu + yv = 0 \tag{65}$$

and

$$u^2 + v^2 - T - y = 0, \tag{66}$$

and the differantial equation

$$T' = -3v \tag{67}$$

The five equations (60) to (63) and (67) are the underlying differential system that contain the solutions of the original problem. However, to impose the three constraints (64) to (66) we must add the term $G^T\mu$ to the differential system. In this case, G is given by

$$G = \begin{pmatrix} 2x & 2y & 0 & 0 & 0 \\ u & v & x & y & 0 \\ 0 & -1 & 2u & 2v & -1 \end{pmatrix} \tag{68}$$

so that the resulting semi-explicit system of ODEs is

$$x' = u + 2x\mu_1 + u\mu_2 \tag{69}$$

$$y' = v + 2y\mu_1 + v\mu_2 - \mu_3 \tag{70}$$

$$u' = -Tx + x\mu_2 + 2u\mu_3 \qquad (71)$$

$$v' = -Ty - 1 + y\mu_2 + 2v\mu_3 \qquad (72)$$

$$T' = -3v - \mu_3 \qquad (73)$$

These, together with eqs (64) - (66), have the same solutions as the original problem. They are an index two, semi-explicit system. The "algebraic" variables μ are zero, so that general implicit multistep methods discussed in Gear [3] can be used. Note that we can get a much simpler index two system consisting of (69) - (72) with μ_2 and μ_3 set to zero and (64) - (65) by differentiating only once. Since variable-step, variable-order BDF methods are known to work for semi-explicit index two systems [5], this is probably a preferable approach for the Euler Lagrange equations.

REFERENCES

[1] BRENAN, K. "Stability and convergence of difference approximations for higher index differential/algebraic equations with applications to trajectory control". PhD Disertation, Math. Dept. UCLA, Los Angeles, 1983.

[2] ENRIGHT, W.H., HULL, T.E., LINDBERG, B. "Comparing Numerical Methods for Stiff Systems of ODEs". BIT, **15** pp. 10-49, 1975.

[3] GEAR, C.W. "Maintaining solution invariants in the numerical solution of ODEs". SIAM J. of Sci. and Stat. Computing (July, 1986), **7**, pp. 734-743.

[4] GEAR, C.W. "Differential-Algebraic Equation Index Transformations". Dept. of Computer Science, Report 1314, University of Illinois at Urbana-Champaign, Dec. 1986. To appear: SIAM J. Sci and Stat. Computing.

[5] GEAR, C.W., GUPTA, G.K. and LEIMKUHLER, B. "Automatic integration of Euler-Lagrange equations with constraints". Journal of Computational and Applied Mathematics (1985), **12-13**, pp. 77-90.

[6] GEAR, C.W., PETZOLD, L.R. "ODE Methods for the Solution of Differential/Algebraic Equations". SIAM Journ. of Numerical Analysis **21**, pp. 716-728, August, 1985.

[7] LOTSTEDT, P. and PETZOLD, L.P. "Numerical solution of nonlinear differential equations with algebraic constraints I: convergence results for backward differentiation formulas". Math Comp (Apr. 1986), **46**, No. 174, pp. 491-516.

[8] MÄRZ, R. "Multistep Methods for Initial Value Problems in Implicit Differential Equations". Humboldt-Universitat zu Berlin, Preprint No. 22, 1981.

[9] PETZOLD, L.P. and LOTSTEDT, P. "Numerical solution of nonlinear differential equations with algebraic constraints II: practical implications". SIAM J. Sci and Stat. Computing (July, 1986), **7**, No. 3, pp. 720-733.

[10] SHAMPINE, L.F. "Conservation Laws and the Numerical Solution of ODEs". Sandia Albuquerque Report SAND84-1241, June 1984.

A COMPARATIVE STUDY OF CHEBYSHEV ACCELERATION AND RESIDUE SMOOTHING IN THE SOLUTION OF NONLINEAR ELLIPTIC DIFFERENCE EQUATIONS

P.J. VAN DER HOUWEN & B.P. SOMMEIJER
Centre for Mathematics and Computer Science
P.O. Box 4079, 1009 AB Amsterdam, The Netherlands

G. PONTRELLI(*)
Istituto per le applicazioni del calcolo
Viale del Policlinico 137, Rome, Italy 00161

Abstract. We compare the traditional and widely-used Chebyshev acceleration method with an acceleration technique based on residue smoothing. Both acceleration methods can be applied to a variety of function iteration methods and allow therefore a fair comparison. The effect of residue smoothing is that the spectral radius of the Jacobian matrix associated with the system of equations can be reduced substantially, so that the eigenvalues of the iteration matrix of the iteration method used are considerably decreased. Comparitive experiments clearly indicate that residue smoothing is superior to Chebyshev acceleration. For a model problem we show that the rate of convergence of the smoothed Jacobi process is comparable with that of ADI methods.
The smoothing matrices by which the residue smoothing is achieved, allow for a very efficient implementation, thus hardly increasing the computational effort of the iteration process. Another feature of residue smoothing is that it is directly applicable to nonlinear problems without affecting the algorithmic complexity. Moreover, the simplicity of the method offers excellent prospects for execution on vector and parallel computers.

1. FUNCTION ITERATION METHODS

In [2] a Jacobi-type iteration method for solving nonlinear elliptic difference equations $\mathbf{f}(\mathbf{u})=\mathbf{0}$ is described which is essentially based on function evaluation without requiring the solution of linear systems during the successive iterations. The

(*) The investigations were supported by the National Research Council (CNR) of Italy.

function values to be evaluated are smoothed residue values **Sf**, where S is a smoothing matrix. This function iteration method (smoothed Jacobi iteration method) is extremely simple to implement on a computer and highly vectorizable on vector computers. The numerical experiments reported in [2] show that smoothed Jacobi iteration is many times faster than conventional Jacobi iteration, indicating that it may be a competitor to other, more sophisticated, function iteration methods for solving nonlinear elliptic difference equations. It is the purpose of this paper to show that smoothed Jacobi iteration is really faster than function iteration methods with a comparable algorithmic complexity. As a reference method we chose the Chebyshev acceleration method applied to Jacobi iteration with automatic estimation of the dominant eigenvalue in order to provide the eigenvalue interval of the Jacobian matrix $\partial \mathbf{f}/\partial \mathbf{u}$ needed by the method. Like smoothed Jacobi iteration, Chebyshev-accelerated Jacobi iteration vectorizes well on vector computers. However, its implementation is more complicated and it turns out that smoothed Jacobi iteration is much faster both for linear and nonlinear problems.

We have tried to accelerate Chebyshev-accelerated Jacobi iteration by applying a technique for eliminating dominant eigenvectors from the iteration error. Since the dominant eigenvalue is automatically determined, such an elimination technique does not complicate the method further. Although we found a reduction of the number of iterations compared with the Chebyshev method without elimination, smoothed Jacobi iteration is still markedly faster.

We then tried to improve the smoothed Jacobi iteration method by applying Chebyshev acceleration, to obtain Chebyshev-accelerated smoothed Jacobi iteration. The results were disappointing. The generally small reduction of the number of iterations does not justify the increased implementational complexity.

Finally, we investigated whether it pays to replace Jacobi iteration by SSOR iteration, to obtain smoothed SSOR iteration. This method requires the evaluation of the Jacobian matrix and is therefore not a true function iteration method any more. Consequently, the convergence improvement should be sufficiently large in order to justify the increased complexity of the method. We found that, when compared with the smoothed Jacobi method, the smoothed SSOR iteration (provided with optimal relaxation parameters) is slighter faster; however, the price to be paid seems not worth the additional effort, and we refrained from a comparison of the Chebychev-accelerated SSOR and the smoothed SSOR methods.

Our conclusion is that smoothed Jacobi iteration is an extremely attractive and efficient method, particularly on vector computers, and that nonlinearities in the system to be solved neither destroy the high rate of convergence, nor increase the algorithmic complexity. We do not claim that this method is faster than, e.g., multigrid methods. However, such methods, even when the underlying relaxation method is based on function iteration, require considerably more implementational effort and are less vectorizable than smoothed Jacobi iteration. In a forthcoming paper we will report on a performance evaluation of smoothed Jacobi iteration on vector computers.

In the remainder of this section we shall briefly describe Chebyshev acceleration with automatic eigenvalue estimation and elimination of dominant eigenvector components in the iteration error, and the idea of residue smoothing. In the Sections 2 and 3 we illustrate these techniques for a few linear and nonlinear examples.

1.1. Chebyshev acceleration

Consider a stationary, linearly convergent one-step iteration method

(1.1) $$\mathbf{u}_{n+1} = \mathbf{F}(\mathbf{u}_n), \quad n \geq 0,$$

where $\mathbf{F}(\mathbf{u}) = \mathbf{u}$, for finding the solution $\mathbf{u}$ of the equation

(1.2) $$\mathbf{f}(\mathbf{u}) = \mathbf{0}.$$

It is explicitly assumed that the iteration function $\mathbf{F}$ does not depend on n, and that $\partial\mathbf{F}/\partial\mathbf{u}$ essentially has a real eigenvalue spectrum. By applying the well-known Chebyshev acceleration method to (1.1) we obtain the two-step semi-iterative method (cf. ,e.g., [1])

(1.3) $$\mathbf{v}_{n+1} = p_n\, \mathbf{v}_n + q_n\, \mathbf{F}(\mathbf{v}_n) + r_n\, \mathbf{v}_{n-1}\,, \quad n \geq 0,$$

where the coefficients are defined by:

(1.4) $$\begin{aligned} p_0 &= w_0 / (w_0 + w_1), \quad p_n = 2w_0\, T_n(w_0+w_1) / T_{n+1}(w_0+w_1), \quad n \geq 1; \\ q_0 &= 1 - p_0, \quad q_n = 2w_1\, T_n(w_0+w_1) / T_{n+1}(w_0+w_1), \quad n \geq 1; \\ r_0 &= 0, \quad r_n = 1 - p_n - q_n, \quad n \geq 1. \end{aligned}$$

Here, T_n denotes the first-kind Chebyshev polynomial of degree n and

(1.5) $$w_0 := -(b + a) / (b - a), \quad w_1 := 2 / (b - a),$$

where, usually, [a,b] denotes the eigenvalue interval of the Jacobian matrix $\partial\mathbf{F}/\partial\mathbf{u}$. One of the end points, say a, of the eigenvalue interval corresponds to the spectral radius of $\partial\mathbf{F}/\partial\mathbf{u}$, and can be estimated by Gerschgorin's disk theorem. The estimation of b is discussed in the next subsection.

1.2. Dominant eigenvectors

In this subsection, we discuss the estimation of the eigenvalue corresponding to the dominant eigenvector. This dominant eigenvalue will provide us with an estimate of b. Furthermore, we consider the elimination of the dominant eigenvector in an attempt to speed up the Chebyshev acceleration process.

1.2.1. Estimation of dominant eigenvalues

Introducing the iteration error

(1.6) $\varepsilon_n = \mathbf{v}_n - \mathbf{u}$,

and substitution into (1.3) yields, in first approximation,

(1.7) $\varepsilon_n \approx P_n(\partial\mathbf{F}/\partial\mathbf{u})\, \varepsilon_0$, $P_n(x) := T_n(w_0 + w_1x) / T_n(w_0 + w_1)$,

where $\partial\mathbf{F}/\partial\mathbf{u}$ denotes the Jacobian matrix of **F** evaluated at the solution **u**; we shall assume that $\partial\mathbf{F}/\partial\mathbf{u}$ exists in the neighbourhood of **u** and does not vanish, and that the initial approximation is already close enough to the true solution, so that second order terms can be neglected.

Suppose that we choose the interval [a,b] such that all eigenvalues of $\partial\mathbf{F}/\partial\mathbf{u}$ are in [a,b] except for one eigenvalue $\lambda^* > b$, with eigenvector $\mathbf{e}^*$. Then, for sufficiently large n, all eigenvector components occurring in the eigenvector expansion of the initial iteration error will be significantly reduced in magnitude, except for the eigenvector $\mathbf{e}^*$; this eigenvector component will dominate the iteration error, i.e.,

(1.8) $\varepsilon_n \approx P_n(\lambda^*)\, \mathbf{e}^*$.

We shall call the above procedure where all eigenvectors but one are reduced in magnitude the *reduction phase* of the iteration method.

From relation (1.8), an estimate for the eigenvalue λ^* can be derived (we shall call λ^* the *dominant eigenvalue*). In the following, division of vectors is always understood to be carried out componentwise.

Theorem 1.1. *Let* w_0, w_1 *be defined by* (1.5), *and let the vector* **R** *be defined by*

$$\mathbf{R} := (w_0 + w_1 + [(w_0 + w_1)^2 - 1]^{1/2})\Delta\, \mathbf{v}_n / \Delta\, \mathbf{v}_{n-1},$$

where Δ *denotes the forward difference operator. Then, a vector of* λ^* *values is provided by*

$$\lambda^* \approx [\mathbf{R}/2 + 1/(2\mathbf{R}) - w_0\mathbf{1}]/w_1,$$

where **1** *denotes the unit vector* $(1, 1, \ldots, 1)^T$.

Proof. From the definition of Chebyshev polynomials we derive, for $\lambda^* > b$,

(1.9) $P_n(\lambda^*) \approx [W(\lambda^*)/W(1)]^n$, $W(x) := w_0 + w_1x + [(w_0 + w_1x)^2 - 1]^{1/2}$.

By rewriting (1.8) for n-1 and n+1, forming the expression

$$[\varepsilon_{n+1} - \varepsilon_n\,] / [\,\varepsilon_n - \varepsilon_{n-1}],$$

and by using (1.9), we find that λ^* approximately satisfies the relation

(1.10) $W(\lambda^*)\,\mathbf{1} = W(1)\,\Delta\,\mathbf{v}_n / \Delta\,\mathbf{v}_{n-1} = \mathbf{R}$.

Solving relation (1.10) for λ^* yields the estimate given in the theorem. ∎

When this theorem is applied in actual computation, we obtain as many estimates to the dominant eigenvalue λ^* as there are equations. The spread $\Delta\lambda^*$ of the interval $[\lambda^*-\Delta\lambda^*, \lambda^*+\Delta\lambda^*]$ of these estimates can be used as an indication to what extent the iteration error is indeed dominated by $\mathbf{e}^*$. The actually used approximation to the dominant eigenvalue might be the arithmetic mean of the available estimates.

1.2.2. Elimination of dominant eigenvectors

Having found an approximation to the dominant eigenvalue during the reduction phase of the iteration method, we can proceed with the elimination of the corresponding eigenvector $\mathbf{e}^*$ from the iteration error (we shall call this process the *elimination phase* of the iterative method). Below, we briefly discuss a few possibilities for eliminating dominant eigenvectors.

One possibility is to apply again the Chebyshev acceleration process (1.3) - (1.4) with the last computed iterate as new initial approximation and with modified values for the parameters w_0 and w_1.

Theorem 1.2. *Let in* (1.4) *the parameters* w_0 *and* w_1 *be defined by*

(1.11) $w_0 = [a \cos(\pi/(2n)) + \lambda^*] / [a-\lambda^*]$, $w_1 = (\cos(\pi/(2n)) + 1) / (\lambda^* - a)$, $a < \lambda^*$,

where λ^* *is the eigenvalue of* $\partial\mathbf{F}/\partial\mathbf{u}$ *corresponding to the eigenvector* $\mathbf{e}^*$ *dominating the iteration error* ε_0.

(a) *Then the Chebyshev method* (1.3) - (1.4) *eliminates* $\mathbf{e}^*$ *from the iteration error after exactly* n *iterations.*

(b) *If the number of iterations is sufficiently large, i.e., if*

(1.12) $n \geq \pi\, [2 \arccos([2\lambda^* - a - 1] / [1 - a])]^{-1}$,

then the method is stable in the sense that no eigenvector components of the iteration error are amplified.

Proof. The expressions (1.11) immediately follow from the conditions that the polynomial $P_n(x)$ should satisfy the relations:

$$P_n(a) = \pm\, 1/ T_n(w_0 + w_1), \quad P_n(\lambda^*) = 0.$$

From these requirements we deduce

$$w_0 + w_1\, a = -1, \quad w_0 + w_1\, \lambda^* = \cos(\pi/(2n)),$$

resulting in (1.11).

The stability condition (1.12) follows from the requirement $w_0 + w_1 \geq 1$. ■

A disadvantage of the above elimination procedure is the computational effort involved by forming a new set of iterates. This leads us to a procedure based on iterates already computed during the reduction phase of the iteration method.

Consider the k+1 iterates $\mathbf{v}_{n-j+1}$, $j = 0, \ldots, k$ computed by (1.3), and define

(1.13) $\mathbf{v}^* = Q(E)\, \mathbf{v}_{n-k+1}$,

where E is the forward shift operator and Q is a polynomial of degree k satisfying the condition Q(1) = 1. For this k-step extrapolation formula the following theorem holds:

Theorem 1.3. *Let* A *be the matrix defined by* $A := \mathrm{diag}\,(\Delta\, \mathbf{v}_n/\Delta\, \mathbf{v}_{n-1})$, *and define* $\varepsilon^* := \mathbf{v}^* - \mathbf{u}$; *then*

$$\|\varepsilon^*\|_2 \le \rho(Q(A))\ \|\varepsilon_{n-k+1}\|_2,$$

where $\|\ \|_2$ *denotes the spectral norm and* ρ *the spectral radius.*

Proof. Using Q(1)=1 we find from (1.13) that

(1.14) $\varepsilon^* = \mathbf{v}^* - \mathbf{u} = Q(E)\, \mathbf{v}_{n-k+1} - \mathbf{u} = Q(E)\, \mathbf{v}_{n-k+1} - Q(E)\, \mathbf{u} = Q(E)\, \varepsilon_{n-k+1}$,

and using (1.7),

(1.15) $\varepsilon^* \approx Q(E)\, P_{n-k+1}(\partial \mathbf{F}/\partial \mathbf{u})\, \varepsilon_0$.

If $\mathbf{e}^*$ dominates the iteration error, then it follows from (1.8) that

$$\varepsilon^* \approx Q(E)\, P_{n-k+1}\,(\lambda^*)\, \mathbf{e}^*,$$

so that, by virtue of (1.9) , we obtain

$$\varepsilon^* \approx [W(\lambda^*)/W(1)]^{n-k+1}\, Q(W(\lambda^*)/W(1))\mathbf{e}^* \approx P_{n-k+1}\,(\lambda^*)\, Q(W(\lambda^*)/W(1))\mathbf{e}^*.$$

Again using (1.8), and replacing $W(\lambda^*)/W(1)$ by a diagonal matrix A with elements defined by the components of the vector $\Delta\, \mathbf{v}_n/\Delta\, \mathbf{v}_{n-1}$, we arrive at the relation

$$\varepsilon^* \approx Q(A)\ \varepsilon_{n-k+1},$$

The assertion of the theorem is now immediate. ■

This theorem suggests that we should choose Q such that its magnitude is small in the interval

(1.16) $[\alpha^*-\Delta\alpha^*, \alpha^*+\Delta\alpha^*] := [\min\{\Delta\, \mathbf{v}_n/\Delta\, \mathbf{v}_{n-1}\}, \max\{\Delta\, \mathbf{v}_n/\Delta\, \mathbf{v}_{n-1}\}]$.

We remark that, if k=1, then, by requiring $Q(\alpha^*)=0$, we obtain the famous one-step extrapolation formula of Lyusternik [4].

It follows from the recursion (1.14) that the extrapolation formula (1.13) is stable if the characteristic polynomial $x^{k+1} - Q(x)$ has its roots on the unit disk, those on the unit circle being simple. It can be shown that, for formulas based on only a few back iterates and for α^* close to 1, this requirement is easily violated if we require at the same time that Q is small in magnitude in the interval (1.16). More stable formulas can be constructed by increasing k. However, this means that more storage is needed to store the necessary iterates.

In actual computation, instabilities introduced by a possible unstable extrapolation formula (1.13) are usually compensated by "overstability" of the reduction phase. To be more precise, we consider the eigenvalues of the amplification matrix occurring in (1.15), i.e. the matrix

$$Q(E)P_{n-k+1}(\partial \mathbf{F}/\partial \mathbf{u}) = Q(E)[T_{n-k+1}(w_0 + w_1\, \partial \mathbf{F}/\partial \mathbf{u})/T_{n-k+1}(w_0 + w_1)].$$

Let Q^* be the polynomial obtained from Q by replacing all the coefficients of Q by their absolute values. Then, all eigenvalues α of this amplification matrix corresponding to eigenvalues of $\partial \mathbf{F}/\partial \mathbf{u}$ lying in [a,b] satisfy the inequality

$$|\alpha| \le Q^*(E)[1/T_{n-k+1}(w_0 + w_1)] \approx 2\, w^{-(n-k+1)}\, Q^*(1/w),\ w := W(1),$$

where W(1) is defined as in (1.9). From this inequality we deduce that the eigenvector components of the initial error are certainly not amplified at the end of the reduction/elimination phase if

(1.17) $$n \ge k - 1 + \log_w(2\, Q^*(1/w)).$$

The following theorem presents the lower bound on n obtained when this result is applied to the case where Q has all its zeros at α^*.

Theorem 1.4. *Let* Q *be given by*

$$Q(x) = [(x - \alpha^*)/(1 - \alpha^*)]^k,$$

where α^* *is defined in* (1.16) *and is assumed to be less than* 1.*Then, at the end of the reduction/elimination phase, no eigenvector components of the initial error are amplified if*

$$n \ge k \log_w((1 + w|\alpha^*|) / (1 - \alpha^*)) + \log_w(2) - 1,\ \ w := W(1),$$

where W(1) *is defined in* (1.9).

Proof. It follows from the definition of Q that

$$Q^*(x) = [(x + |\alpha^*|) / (1 - \alpha^*)]^k.$$

Substitution into (1.17) yields the lower bound on n stated in the theorem. ■

In our experiments, we employed two-step extrapolation formulas because we need already three iterates for estimating the dominant eigenvalue.

1.3. Residue smoothing

In [2] iteration methods employing residue smoothing have been analysed for solving nonlinear elliptic systems $\mathbf{f}(\mathbf{u}) = \mathbf{0}$. Here, we apply the same technique to a more general class of iteration methods. Given a difference matrix D, a set of nonnegative integers $r \in \mathbb{R}$, a set of relaxation parameters ω_r, and some basic iteration method with iteration function

$$\mathbf{G}(\mathbf{u}) := \mathbf{u} + M\ \mathbf{f}(\mathbf{u}),$$

where M is the characterizing matrix. Then, we define the class of smoothed iteration methods

(1.18a) $\quad \mathbf{u}_{n+1} = \mathbf{F}(\mathbf{u}_n), \quad n \geq 0; \quad \mathbf{F} := \prod_{r \in \mathbb{R}} \mathbf{G}_r,$

(1.18b) $\quad \mathbf{G}_r(\mathbf{u}) := \mathbf{u} + \omega_r M \; S_r \, \mathbf{f}(\mathbf{u}),$

(1.18c) $\quad S_r := 4^{-r} \, [T_{2r}(I+2D) - I] \, (2D)^{-1}, \quad T_m(x) := \cos(m \arccos(x)),$

where S_r is the so-called *smoothing matrix* (notice that S_r reduces to the identity matrix if r=0).

Thus, the iteration formula (1.18) may be interpreted as a cycle of smoothed basic iteration steps.

1.3.1. The smoothing matrix

The smoothing matrix S_r is a *polynomial* in D and completely defined as soon as D is specified. We remark that D is allowed to be a singular matrix because, in spite of our notation, its inverse does not need to exist. In order to simplify the notation we shall continue to write D^{-1} in the various formulas without actually requiring that it exists.

The matrix D is a difference matrix the eigenvalues of which are assumed to be in the interval [-1,0]. As a consequence, the eigenvalues of the smoothing matrices are in the interval [0,1]. In one-dimensional problems (two-point boundary value problems), one may consider the difference matrix

$$(1.19) \qquad D := (1/4) \begin{pmatrix} 0 & & & & \\ 1 & -2 & 1 & & \\ & \ddots & \ddots & \ddots & \\ & & 1 & -2 & 1 \\ & & & & 0 \end{pmatrix}$$

This matrix generates smoothing matrices which leave the first and last component of the vector to which they are applied unchanged. Therefore, they are suitable in cases where the first and last component of the vector to be smoothed should not change. For example, we mention the case of a residue vector of which the first and last component vanish, that is, the case where the first and last equation of the system $\mathbf{f}(\mathbf{u}) = \mathbf{0}$ represent *Dirichlet boundary conditions* (we observe that in such situations the first element of the first row and the last element of the last row in D may be replaced by any value, so that D becomes nonsingular).

In actual computation, it is generally not feasible to precompute the smoothing matrix because of storage requirements. On the other hand, in the case where D is defined by (1.19), the matrix S_r exhibits a regular pattern which can be exploited for a an efficient implementation. For example, the first three smoothing matrices are respectively given by S_0=I,

$$S_1 = (1/4)\begin{pmatrix} 4 & & & & & & \\ 1 & 2 & 1 & & & & \\ & & \cdot & \cdot & \cdot & & \\ & & & \cdot & \cdot & \cdot & \\ & & & & 1 & 2 & 1 \\ & & & & & & 4 \end{pmatrix}$$

and

$$S_2 = (1/16)\begin{pmatrix} 16 & & & & & & & & \\ 9 & 2 & 2 & 2 & 1 & & & & \\ 4 & 2 & 4 & 3 & 2 & 1 & & & \\ 1 & 2 & 3 & 4 & 3 & 2 & 1 & & \\ 0 & 1 & 2 & 3 & 4 & 3 & 2 & 1 & \\ 0 & 0 & 1 & 2 & 3 & 4 & 3 & 2 & 1 \\ & & & & \cdot & \cdot & \cdot & & \end{pmatrix}$$

These examples suggest the precomputation of the first few rows of the smoothing matrices until the pattern becomes regular. Alternatively, we can generate the smoothing matrices by exploiting factorization properties of Chebyshev polynomials. This approach seems to be more attractive. By using the identity

$$T_{2^{r+1}}(z) \equiv T_2(T_{2^r}(z)),$$

we derive from (1.18c) the recursion

$$S_{r+1} = (I + 4^r D\, S_r)\, S_r.$$

By writing $S_{r+1} = F_{r+1} S_r$ we arrive at the following theorem which expresses the smoothing matrix S_r as a product of r factor matrices F_j:

Theorem 1.5. *Let* D *be any difference matrix and define* the factor matrices $F_1 := I + D$, $F_{j+1} := (I - 2F_j)^2, j \geq 1$. *Then* $S_r = F_1 . F_2 . \ldots . F_r$. ■

From this theorem we conclude that, if the factor matrices are precomputed, then the smoothing matrix can be generated by r matrix-vector multiplications. For a number of difference matrices D and for r not too large (in a typical case r should not exceed $\log_2(1/\Delta x)$ where Δx is the mesh size), it turns out that the corresponding factor matrices are almost as sparse as D itself, so that the application of the smoothing matrix involves only r matrix-vector multiplications with matrices of similar complexity as D. For larger values of r one may proceed as follows. Let S_q be the smoothing matrix which can 'conveniently' be generated by means of Theorem 1.5. Then, by employing the factorization formula

$$T_{2^{q+j}}(z) \equiv T_{2^j}(T_{2^q}(z)),$$

we find that

$$S_{q+j} := 4^{-q-j}\, [T_{2^j}(I + 2^{2q+1} D\, S_q\,) - I]\, (2D)^{-1}.$$

This formula expresses S_{q+j} in terms of a polynomial of degree at most 2^j of D and S_q. In this way it is in principle possible to generate smoothing matrices of higher indices.

The precomputation of the factor matrices is relatively easy in the case of *one-dimensional* problems and can often be done by hand. In higher dimensional elliptic problems with irregular geometries, this is less attractive. However, by considering the problem as a system of coupled two-point boundary-value problems, one may apply one-dimensional smoothing operators (e.g., based on (1.19)) to the successive problems (cf. [2]).

1.3.2. The matrix M

By means of the matrix M, several basic iteration methods can be selected. Let the matrix $\partial\mathbf{f}/\partial\mathbf{u}$ be split according to

$$\partial\mathbf{f}/\partial\mathbf{u} = C + L + U \quad \text{or} \quad \partial\mathbf{f}/\partial\mathbf{u} = H + V,$$

where C is diagonal, L and U are lower and upper triangular, and H, V correspond to the ADI splitting of A (or any other 'convenient' splitting). In terms of these splitting matrices, various matrices M may be defined. In Table 1.1, a few examples are listed. Here, ϖ , ϖ_H and ϖ_V are parameters which depend on the spectrum of $\partial\mathbf{f}/\partial\mathbf{u}$ (cf. [1]).

Table 1.1. Possible M matrices.

Jacobi-type:	$M :=$ diagonal, e.g., $M = - C^{-1}$
Gauss-Seidel:	$M := - (C + L)^{-1}$
SOR:	$M := - (C/\varpi + L)^{-1}$
SSOR:	$M := - (2/\varpi - 1)\, (C/\varpi + U)^{-1}\, C\, (C/\varpi + L)^{-1}$
ADI:	$M := - (\varpi_H + \varpi_V)\, (V + \varpi_V I)^{-1}\, (H + \varpi_H I)^{-1}$

We observe that instead of solving $\mathbf{f}(\mathbf{u}) = \mathbf{0}$, we can alternatively solve the preconditioned system $P\mathbf{f}(\mathbf{u}) = \mathbf{0}$. If we replace the matrix MS_r by MS_rP, then the resulting iteration method is given by (1.18). In particular, we may set M = I and P equal to one of the matrices specified above.

1.3.3. The model situation

The choice of the relaxation parameters in (1.18) will be based on the model case where D equals the difference matrix defined by the normalized Jacobian of **f**:

(1.20) $$D := \rho^{-1}\, \partial \mathbf{f}/\partial \mathbf{u},$$

with ρ denoting the spectral radius of $\partial \mathbf{f}/\partial \mathbf{u}$. However, we emphasize that the matrix D actually used in practice is a very rough approximation to this normalized Jacobian; for example, the matrix defined in (1.19) turned out to be rather effective in the case of smoothed Jacobi iteration of Dirichlet problems (cf. [2]). The damping of the iteration error, is largely determined by the Jacobian matrix $\partial \mathbf{F}/\partial \mathbf{u}$ in (1.18):

(1.21) $$\partial \mathbf{F}/\partial \mathbf{u} = \prod_r \partial \mathbf{G}_r/\partial \mathbf{u} = \prod_r \,[I + \omega_r M S_r\, \partial \mathbf{f}/\partial \mathbf{u}].$$

Substitution of (1.20) into (1.18c) and the resulting expression for S_r into (1.21) yields

(1.22) $$\partial \mathbf{F}/\partial \mathbf{u} = \prod_r [I + \rho\, \omega_r\, 4^{-r} M\, [T_{2r}(I+2D) - I]/2].$$

Thus, given the matrices M and D, we are faced with the problem of choosing a set of relaxation parameters $\{\omega_r\}$ such that the eigenvalues of $\partial \mathbf{F}/\partial \mathbf{u}$ are small in magnitude. These eigenvalues will be called *damping factors* of the iteration method. In Section 3 we will derive suitable relaxation parameters for the Jacobi case. The resulting iteration scheme belongs to the class of function iteration methods, which, essentially, only require the evaluation of values of **f**. In Section 4 we describe a numerical approach to obtain relaxation parameters for the SSOR method. Formally, these schemes do not belong to the class of function iteration methods, although the amount of linear algebra for these schemes is rather modest. Still further away from this class is the ADI case, which involves the solution of tridiagonal systems. We did not consider this method; an analysis of smoothed ADI iteration may be found in [5].

First, however, for the sake of comparison, we give results obtained by Chebyshev-accelerated Jacobi iteration which still belongs to the most efficient conventional function iteration methods available in the literature for solving elliptic equations.

2. CHEBYSHEV ACCELERATION OF JACOBI ITERATION

We shall present numerical experiments with the Chebyshev acceleration method of conventional Jacobi-type iteration with automatic estimation of the dominant eigenvalue. The most simple choice of the matrix M characterizing a Jacobi-type iteration method is $M = 2I/\rho$, where ρ is the spectral radius of $\partial \mathbf{f}/\partial \mathbf{u}$.

Alternatively, one may choose $M = -[\mathrm{diag}(\partial \mathbf{f}/\partial \mathbf{u})]^{-1} = -C^{-1}$. We emphasize that, updating the matrix M during the iteration may result in an iteration function **F** which is n-dependent, contradicting our assumption that **F** is stationary (see Section 1.1). If no Chebyshev acceleration is applied, then, at the cost of some additional computational effort, this strategy may improve the convergence (see Section 3.2). However, if Chebyshev acceleration is used, then the matrix M should be evaluated in the first step of the reduction phase.

First, we apply the Chebyshev-accelerated Jacobi method without elimination of the dominant eigenvectors. In the next subsection we shall illustrate the effect of the elimination process.

2.1. Chebyshev acceleration with automatic estimation of the dominant eigenvalue

The following strategy was applied:

(i) *Initial approximation*: linear interpolation of the boundary values.

(ii) *Chebyshev reduction phase*: application of the Chebyshev acceleration process {(1.3) - (1.5)} where **F** is defined by (1.18) with $\mathbb{R}=\{0\}$, $\omega_0=1/2$, a=0, and where the value of b occurring in {(1.3) - (1.5)} is such that the dominant eigenvalue λ^* is outside the interval [a,b] (observe that this choice of $\mathbb{R}$ results in a conventional iteration method because the only smoothing matrix $S_r = S_0 = I$). In our experiments we chose b=0.95.

(iii) *Restart criterion*: restart of the reduction phase with adjusted value of b as soon as the λ^*-estimates obtained in two successive iterations (cf. Theorem 1.1 and the discussion following this theorem) satisfy the condition $\|\lambda^*_n - \lambda^*_{n-1}\| < \eta\, \lambda^*_n$. The new value of b is defined by $b=\lambda^*_n+\delta(1-\lambda^*_n)$. The strategy parameters η and δ are specified in the tables of results.

(iv) *Stopping criterion*: termination of the iteration process as soon as the residue satisfies the condition $\|\mathbf{f}(\mathbf{u}_n)\|_\infty \le 10^{-2}\,(\Delta x)^2$, where Δx denotes the mesh size of the grid defining the elliptic difference equations.

Consider the model problem $u_{xx}= g(x)$ with Dirichlet boundary conditions at x=0 and x=1, that is the system

$$
(2.1)\qquad
\begin{pmatrix}
1 & & & & \\
1 & -2 & 1 & & \\
 & \cdot & \cdot & \cdot & \\
 & & \cdot & \cdot & \cdot \\
 & & 1 & -2 & 1 \\
 & & & & 1
\end{pmatrix}
\begin{pmatrix} u_0 \\ u_1 \\ \cdot \\ \cdot \\ u_m \\ u_{m+1} \end{pmatrix}
=
\begin{pmatrix} u(0) \\ (\Delta x^2) g_1 \\ \cdot \\ \cdot \\ (\Delta x^2) g_m \\ u(1) \end{pmatrix}
$$

where $\Delta x := 1/(m+1)$, $g_j := g(j\Delta x)$, and where $u(0)$ and $u(1)$ are prescribed boundary values.

We start with this model problem where

(2.2) $$g(x) = 6x,\ u(0) = 0 \text{ and } u(1) = 1.$$

For future reference, we first give the results of the conventional Jacobi method for a few values of Δx. The numbers of iterations needed to satisfy the stopping criterion (iv) are given in Table 2.1.

Table 2.1. Conventional Jacobi method for the model problem with $M = (\Delta x)^2 I/2$.

$\Delta x=1/16$	$\Delta x=1/32$	$\Delta x=1/64$
1190	5342	23675

Next, we apply the Chebyshev acceleration, following the above mentioned strategy. For a few values of Δx, η and δ, the numbers of iterations are listed in Table 2.2. The smallest number on each grid is printed in bold type.

Table 2.2. Chebyshev-accelerated Jacobi method for the model problem {(2.1)-(2.2)} with $M = (\Delta x)^2 I/2$

η	$\Delta x=1/16$			$\Delta x=1/32$			$\Delta x=1/64$		
	$\delta=0$	$\delta=.1$	$\delta=.25$	$\delta=0$	$\delta=.1$	$\delta=.25$	$\delta=0$	$\delta=.1$	$\delta=.25$
10^{-2}	159	148	131	658					
10^{-3}	72	67	78	265	242	199	983		
10^{-4}	74	**66**	79	198	160	163	694	643	555
10^{-5}		79	66	166	**158**	175	554	501	**388**
10^{-6}			71	173	161	180	402	394	417

The second example is a nonmodel problem originating from the nonlinear problem

(2.3) $$(\exp(u))_{xx} - 5\,x^3\,(4 + 5u)\exp(u) = 0,\ u(0) = 0,\ u(1) = 1,$$

the exact solution of which is given by $u(x)=x^5$. This problem is discretized on the grid points $\{j\Delta x\}$ using symmetric differences.
Again, we start with the results obtained by the conventional Jacobi method:

Table 2.3. Conventional Jacobi method for problem (2.3) with $M= (\Delta x)^2 I / (2e)$.

$\Delta x=1/16$	$\Delta x=1/64$
1925	38117

The effect of the Chebyshev acceleration on this problem is shown in Table 2.4.

Table 2.4. Chebyshev-accelerated Jacobi method for problem (2.3) with $M = (\Delta x)^2 I / (2e)$.

η	$\Delta x=1/16$			$\Delta x=1/32$			$\Delta x=1/64$		
	$\delta=0$	$\delta=.1$	$\delta=.25$	$\delta=0$	$\delta=.1$	$\delta=.25$	$\delta=0$	$\delta=.1$	$\delta=.25$
10^{-2}	415			1417					
10^{-3}	415	392	356	1222					
10^{-4}	151	137	**98**	1235	1169	1064	2917	2763	2513
10^{-5}	151	137	**98**	1235	1169	1064	2417	2289	2081
10^{-6}	212	191	213	389	359	**278**	1177		**1001**

In performing the above experiments, we observed that the estimates λ^*_n converged from below to the true value. This explains why the best results are obtained by slightly overestimating (i.e., by setting $\delta >0$) the final estimate. Furthermore, the convergence of λ^*_n happened to be very slow. Therefore, it is not surprising that a rather stringent restart criterion (e.g. $\eta \in [10^{-6},10^{-5}]$) results in an optimal performance.

2.2. Chebyshev acceleration with automatic elimination of dominant eigenvectors

Instead of the restart criterion (iii) of the preceding subsection we now use:

(iii) *Elimination phase:* application of a three-step elimination formula of the form (1.13) with characteristic polynomial defined according to Theorem 1.4, and restart of the reduction phase (ii) with adjusted value of b. The elimination formula is applied as soon as (1.17), with k=3, is satisfied, and if the λ^*-estimates obtained in two successive iterations satisfy the condition $|\lambda^*_n - \lambda^*_{n-1}| < \eta\, \lambda^*_n$. The new value of b is defined by $b = \gamma \lambda^*_n$, where the strategy parameter γ is specified in the tables of results. The analogues of the Tables 2.2 and 2.4 are given below. The numbers in brackets denote the number of times that a dominant eigenvector has been eliminated. A comparison of the results listed in these tables with those of the Tables 2.2 and 2.4 reveals that the reduction of the number of iterations is rather modest and does not seem worth the additional implementational effort.

Table 2.5. Chebyshev-accelerated Jacobi method for the model problem {(2.1)-(2.2)} with $M = (\Delta x)^2 I / 2$.

η	Δx=1/16		Δx=1/32		Δx=1/64	
	γ=.95	γ=.99	γ=.95	γ=.99	γ=.95	γ=.99
10^{-2}	55(5)	75(3)	141(10)	174(8)	483(23)	438(12)
10^{-3}	52(3)	71(3)	135(7)	188(7)	445(21)	**435(11)**
10^{-4}	**49(2)**	66(2)	212(7)	156(5)	856(21)	456(9)
10^{-5}	**49(2)**	57(1)	192(4)	135(3)	1074(10)	477(7)
10^{-6}	62(2)	56(1)	209(3)	**129(3)**		

Table 2.6. Chebyshev-accelerated Jacobi method for problem (2.3) with $M = (\Delta x)^2 I / (2e)$.

η	Δx=1/16		Δx=1/32		Δx=1/64	
	γ=.95	γ=.99	γ=.95	γ=.99	γ=.95	γ=.99
10^{-2}	96(6)	124(5)	303(16)	265(8)	**611(31)**	622(14)
10^{-3}	137(6)	121(4)	298(14)	**261(8)**	656(30)	640(14)
10^{-4}	109(4)	100(3)	463(14)	263(6)	1258(31)	673(14)
10^{-5}	130(4)	**94(2)**	590(6)	392(4)	1963(19)	1025(13)
10^{-6}	130(4)	183(1)	491(4)	401(3)		

3. SMOOTHED JACOBI ITERATION

As we have seen, if M is chosen to be a *diagonal* matrix, then the basic iteration method is of the Jacobi-type. In the previous section, we selected $\mathbb{R} = \{0\}$, resulting in $S_r = I$, i.e., an unsmoothed process. In this section we shall exploit the matrix S_r, that is we consider a *smoothed* Jacobi-type iteration method.

We shall first derive suitable relaxation parameters for the model situation (1.20), and then we shall show, by means of numerical experiments, that these parameters are also effective in nonmodel cases.

3.1. Derivation of relaxation parameters

Let M be the identity matrix (or any diagonal matrix with constant diagonal entries), and let μ, $\alpha_r(\mu)$ and $\alpha(\mu)$ denote the eigenvalues of D, $\partial \mathbf{G}_r/\partial \mathbf{u}$ and $\partial \mathbf{F}/\partial \mathbf{u}$, respectively. An inspection of the zeros and extreme values of the functions $\alpha_r(\mu)$ reveals that, if the set $\mathbb{R}$ contains an integer r, then it should contain the integers 0,... , r-1, otherwise $\alpha(\mu)$ assumes values 1 in the interval [-1,0). This leads us to define the set $\mathbb{R}$ by successive integers starting with r=0. From the expression (1.22) we obtain the following theorems:

Theorem 3.1. *Let* D *and* $\partial \mathbf{f}/\partial \mathbf{u}$ *be related according to* (1.20), *let* $\mathbb{R} := \{0, 1, \ldots, q\}$, *and let* $\rho\omega_r 4^{-r}M = I$ *for all r in* $\mathbb{R}$. *Then the Jacobian matrix* $\partial \mathbf{F}/\partial \mathbf{u}$ *of the iteration function in* (1.18) *and the corresponding damping factors are respectively given by* S_{q+1} *and by*

$$\alpha(\mu) := 2^{-(2q+3)}\, \mu^{-1}[T_{2q+1}(1 + 2\mu)-1],$$

where μ *runs through the eigenvalues of* D.

Proof. On substitution of $\mathbb{R}$ and $\rho\omega_r 4^{-r}M = I$ into (1.22) we obtain

$$\partial \mathbf{F}/\partial \mathbf{u} = \prod_r [I + T_{2^r}(I + 2D)/2].$$

Using a factorization formula for Chebyshev polynomials of degree $m = 2^p$ (cf. [2]):

$$T_m(z) = 1 - m(1-z) \prod_{j=0}^{p-1} (1 + T_{2^j}(z)),$$

we find that $\partial \mathbf{F}/\partial \mathbf{u} = S_{q+1}$ which yields the assertion of the theorem. ■

Theorem 3.2. *Let the conditions of Theorem* 3.1 *be satisfied, let the largest value in the interval* [-1,0)*where the function* $\alpha(\mu)$ *assumes a maximum value be denoted by* μ_b, *and let* μ_s *be the largest value in* [-1,0) *where* $\alpha(\mu) = \alpha(\mu_b)$.*Then the following assertions hold*:

(a) *If the eigenvalues* μ *of* D *satisfy the inequality* $-1 \le \mu \le \mu_s$, *then the spectrum of the matrix* $\partial \mathbf{F}/\partial \mathbf{u}$ *is contained in the interval* $[a,b] := [0, \alpha(\mu_s)]$.
(b) *For all* q *we have the approximation* $\mu_s \approx [\cos(\pi/2q) - 1]/3 \approx -4^{-q}\pi^2/6$.

Proof. It follows from Theorem 3.1 that the eigenvalues of the matrix $\partial \mathbf{F}/\partial \mathbf{u}$ are given by $\alpha(\mu)$, where μ runs through the spectrum of D (see Figure 3.1). From the definition of μ_b and μ_s it follows that

$$0 \le \alpha(\mu) \le \alpha(\mu_b) = \alpha(\mu_s) \text{ for all } \mu \in [-1, \mu_s]$$

proving part (**a**) of the theorem.
A numerical calculation reveals that

$$\mu_s \approx 2\,\mu_z/3,$$

with μ_z the largest value in the interval [-1,0) where the function $\alpha(\mu)$ assumes a zero value. This leads to the approximation given by part (**b**). ■

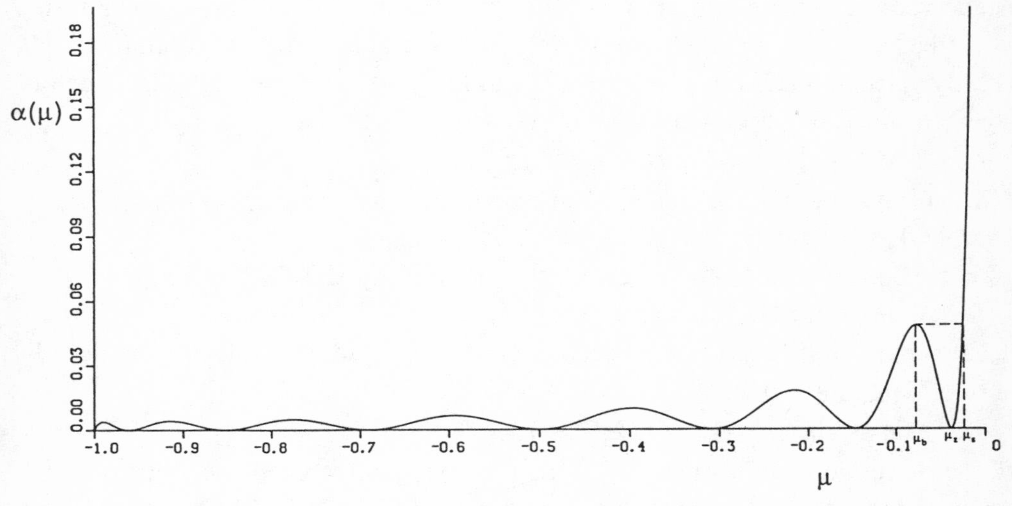

Figure 3.1. Behaviour of the function $\alpha(\mu)$ for q=3

Recalling that the eigenvalues of $\partial \mathbf{F}/\partial \mathbf{u}$ are the damping factors of the iteration method, it is of interest to see to what extent the eigenvalue interval [a,b] = [0,b] of $\partial \mathbf{F}/\partial \mathbf{u}$ is reduced. In Table 3.1 the numerical values of b for a few values of q are given. These values show that for q ≥ 3 this interval is almost constant and approximately given by [0,.05], provided, of course, that the eigenvalues of D are less than μ_s.

Table 3.1. Numerical values of μ_s and $b = \alpha(\mu_s)$

q =	1	2	3	4	5	6	7	8
$-\mu_s$ =	1/3	$.97_{10}{-1}$	$.25_{10}{-1}$	$.64_{10}{-2}$	$.16_{10}{-2}$	$.40_{10}{-3}$	$.99_{10}{-4}$	$.25_{10}{-4}$
b =	.0741	.0525	.0485	.0475	.0473	.0472	.0472	.0472

The following example, illustrates this result.

Example 3.1. Consider the system (2.1) arising from the equation $u_{xx} = g(x)$ with Dirichlet boundary conditions at x=0 and x=1. Let the matrix D be defined by (1.19), then the condition (1.20) is satisfied with $\rho = 4(\Delta x)^{-2}$. By virtue of the Dirichlet boundary conditions incorporated in (2.1), we can restrict the space of residue vectors to the subspace of vectors with vanishing first and last component. Let D* be the matrix obtained by omitting the first and last row and column of D. It is easily verified that, in this subspace, D and D* have the same set of eigenvectors and eigenvalues. It is well known that D* possesses eigenvalues given by

$$\mu_j^* = -[1 - \cos(j\pi/(m+1))]/2, \quad j = 1, \ldots, m,$$

where m is the order of the matrix D*. Thus, the relevant eigenvalues of D are in the interval $(-1, -[1 - \cos(\pi/(m+1))]/2] \approx (-1, -\pi^2/(4(m+1)^2)]$.
A comparison with the bound μ_s given in Theorem 3.2 yields the condition

$$m \leq (3\, 4^q/2)^{1/2} - 1 = 2^q \sqrt{1.5} - 1.$$

On the other hand, in order to preserve a simple structure of the factor matrices F_j we should require that

$$q \leq \log_2(1/\Delta x) \Rightarrow m \geq 2^q - 1.$$

(cf. the discussion of Theorem 1.5). This leads us to the conclusion that smoothed Jacobi iteration has damping factors bounded by .05 if $q \geq 3$ and if m satisfies the above inequalities. For future reference, we list the bounds on m for a few values of q (cf. Table 3.2). ■

Table 3.2. Lower and upper bounds for m for the model problem (2.1).

q =	3	4	5	6	7	8
m≥	7	15	31	63	127	255
m≤	8	18	38	77	155	312

Next we consider the average rate of convergence of smoothed Jacobi for the above model problem, or more generally, for problems which satisfy the conditions of Theorem 3.1 and 3.2 (a). Then the following theorem holds.

Theorem 3.3. *Let the conditions of Theorem* 3.1 *be satisfied, let the eigenvalues* μ *of* D *satisfy the inequality* $-1 \leq \mu \leq \mu_s$, *and let* $q \approx \log_2(1/\Delta x)$ *as* $\Delta x \to 0$, *then the average rate of convergence of smoothed Jacobi iteration is given by* $c/\ln(1/\Delta x)$ *where* $c \approx 2.1$ *as* $\Delta x \to 0$.

Proof. It follows from Table 3.1 that per iteration step the average reduction factor for the iteration error is given by $r := b^{1/(q+1)} \approx .05^{1/(q+1)}$. Hence, the average rate of convergence is given by $R := -\ln(r) \approx 3/(q+1)$, so that for $q \approx \log_2(1/\Delta x)$ we obtain $R \approx 2.1/\ln(1/\Delta x)$.

The condition that q should be as large as $\log_2(1/\Delta x)$ without violating the condition $-1 \leq \mu \leq \mu_s$ can be satisfied in case of the model problem considered in Example 3.1. In the case of two-dimensional model problems, these conditions can also be satisfied provided that we base the smoothing procedure on the successive application of one-dimensional smoothing matrices. In fact, the value of b will be slightly smaller than .05 resulting in a slightly larger value for c. For such problems it is of interest to compare the average rate of convergence of smoothed Jacobi with that of ADI methods. For the Peaceman-Rachford version of the ADI method it is known that the average rate of convergence is given by $R \approx c/\ln(1/\Delta x)$, where c is some constant greater than .777. Thus, we may conclude that smoothed Jacobi has the same order of convergence rate as the ADI method, but is much cheaper per iteration step because of the absence of implicit relations to be solved.

3.2. Numerical experiments

In our numerical experiments, we applied Chebyshev acceleration of smoothed Jacobi iteration with prescribed interval [a,b] according to the following strategy:

(i) *Initial approximation*: linear interpolation of the boundary values.

(ii) *Chebyshev reduction phase*: application of the Chebyshev acceleration process {(1.3) - (1.5)} where **F** is defined by (1.18) and (1.19) with $\omega_r = 2^{2r-1}$ (r=0,...,q). These ω_r parameters give rise to a zero a-value. The value of b is specified in the tables of results and q and m are chosen as allowed by Table 3.2.

(iii) *Stopping criterion*: termination of the iteration process as soon as the residue satisfies the condition $\|\mathbf{f}(\mathbf{u}_n)\|_\infty \leq 10^{-2} (\Delta x)^2$, where Δx denotes the mesh size of the grid on which the elliptic difference equations are defined.

As in the preceding section, possible choices of the matrix M are $M = 2I/\rho$, where ρ is the spectral radius of $\partial\mathbf{f}/\partial\mathbf{u}$, or $M = -[\mathrm{diag}(\partial\mathbf{f}/\partial\mathbf{u})]^{-1}$.

Again we start with the model problem defined by (2.1) with g(x) = 6x, u(0) = 0 and u(1) = 1. In Table 3.3 the numbers of iterations needed to satisfy the stopping criterion are listed. It turns out that if smoothing is used, then the exact solution of the system of equations is obtained after just one cycle of smoothed Jacobi iterations, that is, after one single application of the iteration formula (1.18). This means that the value of b is irrelevant, because the iteration process stops before the Chebyshev recursion gets started. The reason for this peculiar behaviour is that for model problems of the type (2.1) and for the special grids employed in Table 3.3, all damping factors of the smoothed Jacobi method (as specified above) vanish.

Table 3.3. Smoothed Jacobi method for the model problem with $M = (\Delta x)^2 I / 2$.

Δx=1/16	Δx=1/32	Δx=1/64	Δx=1/128	Δx=1/256
5	6	7	8	9

Our second example is the nonmodel problem (2.3). The results of the smoothed Jacobi method without acceleration (first row in Table 3.4) show an impressive reduction of the number of iterations when compared with conventional Jacobi (see Table 2.3). But also the comparison of smoothed Jacobi and Chebyshev-accelerated Jacobi (see the Tables 2.4 and 2.6) clearly shows the superiority of residue smoothing as an acceleration technique. We then tried to improve smoothed Jacobi further by applying Chebyshev acceleration to the smoothed Jacobi process. Table 3.4 indicates only a modest increase of the rate of convergence, especially in cases where the b-value is not optimal. Therefore, one may decide to forget about Chebyshev acceleration in the case of residue smoothing. This results in one array less for storage and at the same time in an extremely simple algorithm.

Table 3.4. Smoothed Jacobi method for problem (2.3) with $M = (\Delta x)^2 I / (2e)$.

Chebyshev acceleration	[a,b]	Δx=1/16	Δx=1/64	Δx=1/256
no	--	73	129	196
yes	[0,.75]	78	137	197
yes	[0,.50]	48	88	125
yes	[0,.40]	**46**	**79**	**117**
yes	[0,.30]	55	96	142
yes	[0,.20]	62	108	162

Our next experiment illustrates the effect of tuning the matrix M to the diagonal of the Jacobian matrix $\partial \mathbf{f}/\partial \mathbf{u}$. Table 3.5 shows that for problem (2.3) some reduction of the number of iterations is obtained, but it is doubtful whether it is worth the additional effort for computing the diagonal elements. A second observation is that for $M = -[\partial \mathbf{f}(\mathbf{u}_n)/\partial \mathbf{u}]^{-1}$ the Chebyshev acceleration does not improve the convergence, because the iteration function **F** is non-stationary.

Table 3.5. Smoothed Jacobi method for problem (2.3) with alternative M matrices.

		$M = -[\partial \mathbf{f}(\mathbf{u}_0)/\partial \mathbf{u}]^{-1}$			$M = -[\partial \mathbf{f}(\mathbf{u}_n)/\partial \mathbf{u}]^{-1}$		
Chebyshev acceleration	[a,b]	$\Delta x=1/16$	$\Delta x=1/64$	$\Delta x=1/256$	$\Delta x=1/16$	$\Delta x=1/64$	$\Delta x=1/256$
no	--	**32**	55	87	**33**	**55**	**80**
yes	[0,.75]	99	179	260	124	216	332
yes	[0,.50]	58	102	151	64	104	169
yes	[0,.40]	48	88	125	54	96	134
yes	[0,.30]	43	74	107	44	82	116
yes	[0,.20]	37	62	95	39	69	98
yes	[0,.10]	**32**	**53**	**79**	34	61	89

4. SMOOTHED SSOR ITERATION

If M is chosen according to the SSOR matrix listed in Table 1.1, then the resulting iteration method becomes a smoothed SSOR iteration method. We shall first derive suitable relaxation parameters for the model situation (1.20), and then we shall show, by means of numerical experiments, that these parameters are also effective in nonmodel cases.

4.1. Derivation of relaxation parameters

Assuming that (1.20) is satisfied, we find, upon substitution of the SSOR matrix in the expression (1.22), the matrix

(4.1) $$\partial \mathbf{F}/\partial \mathbf{u} = \prod_{r \in \mathbb{R}} [I - \rho\, \omega_r 4^{-r} (2/\varpi - 1) (C/\varpi + U)^{-1} C (C/\varpi + L)^{-1} [T_{2^r}(I+2D) - I]/2].$$

In the following we shall allow that ϖ also depends on r, and we shall write $\gamma_r := \omega_r 4^{-r}$.

From this expression we obtain the following theorem:

Theorem 4.1. **(a)** *Let* D *and* $\partial \mathbf{f}/\partial \mathbf{u}$ *be related according to* (1.20). *Then the Jacobian matrix* $\partial \mathbf{F}/\partial \mathbf{u}$ *of the iteration function in* (1.18) *is given by*

(4.2) $$\partial \mathbf{F}/\partial \mathbf{u} = \prod_{r \in \mathbb{R}} [I - \gamma_r (2 - \varpi_r)(D + E)^{-1} [T_{2r}(I + 2D) - I]/2],$$

$$E := [\varpi_r LC^{-1}U + (1-\varpi_r)C/\varpi_r]/\rho$$

(b) *If* $\partial \mathbf{f}/\partial \mathbf{u}$ *is a tridiagonal matrix with lower diagonal, diagonal and upper diagonal elements respectively given by* l_j, j=2, ... , m, *by* c_j, j=1, ... ,m, *and by* u_j, j=1, ... ,m-1, *then* E *is a diagonal matrix with entries*

(4.3) $$e_j = [\varpi_r l_j u_{j-1} / c_{j-1} + (1-\varpi_r) c_j / \varpi_r] / \rho, \quad j = 1, \dots ,m; \; l_1 = 0,$$

with e_0 *and* e_{m+1} *irrelevant.*

Proof. It is easily shown that the matrix (4.1) can be written in the form

(4.1') $$\partial \mathbf{F}/\partial \mathbf{u} = \prod_{r \in \mathbb{R}} [I - \rho\gamma_r (2 - \varpi_r)(\partial \mathbf{f}/\partial \mathbf{u} + \rho E)^{-1} [T_{2r}(I + 2D) - I]/2],$$

where E is defined as in (4.2). Hence, using (1.20) yields the representation (4.2). The proof of part **(b)** of the theorem is straightforward by verification. ■

In order to get some insight into the eigenvalues of the matrix $\partial \mathbf{F}/\partial \mathbf{u}$, we consider the case where $\partial \mathbf{f}/\partial \mathbf{u}$ is tridiagonal; then the theorem states that E is diagonal. The usual approach now is to apply the frozen coefficient technique, that is, to consider the entries of the matrix E to be independent of j, thus simulating the analysis for a linear problem. Unfortunately, even in the case of the model problem, the assumption that E is a constant matrix is not true. This can be seen from the definition of E: all entries of E are equal (in the model case), except for the first element e_1, because the matrix $LC^{-1}U$ gives no contribution for this first element (l_1=0). This exception has some consequences which will be discussed below. For the moment we ignore this deficiency and continue the analysis.

Let us write

$$e_j = d\,\varpi_r + c\,(1-\varpi_r)/\varpi_r, \quad j = 1, \dots , m,$$

with d and c constant (for example, in the model problem (2.1), we have d = - 1/8 and c = - 1/2), then the eigen- values of $\partial \mathbf{F}/\partial \mathbf{u}$ are given by

(4.4) $$\alpha(\mu) = \prod_{r \in \mathbb{R}} \alpha_r(\mu),$$

$$\alpha_r(\mu) := 1 - \gamma_r (2 - \varpi_r) [\mu + d\,\varpi_r + c\,(1-\varpi_r)/\varpi_r]^{-1} [T_{2r}(1 + 2\mu) - 1]/2.$$

Here, as before, μ runs through the eigenvalues of D.

Using this expression, we performed numerically a minimization process for the maximal value of $|\alpha(\mu)|$ on the eigenvalue interval of D. In this minimization process we imposed the constraint $0 < \varpi_r < 2$, which is natural in the SSOR-context. Furthermore, because of the deficiency discussed above, we examined the eigenvalues of $\partial \mathbf{F}/\partial \mathbf{u}$ for the first element of the cycle separately (i.e., for $r = 0$), using the actual value for e_1. That is, we considered the matrix $(D+E)^{-1}D$ (cf. (4.2)). It turned out that, for all values of ϖ_0, this matrix possesses an eigenvalue $1/(2-\varpi_0)$, resulting in an eigenvalue $1-\gamma_0$ for $\partial \mathbf{F}/\partial \mathbf{u}$. Hence, convergence of this unsmoothed SSOR process requires $0 < \gamma_0 < 2$. The necessity of this requirement was experimentally verified. For this reason, we imposed the additional constraint $0 < \gamma_r < 2$ in the minimization process.

The eigenvalues of D defined by (1.19) are known from which we derive $-1 \leq \mu \leq -[1-\cos(\pi/2^q)]/2$, where we have assumed the relation $m + 1 = 2^q$, $m + 2$ being the dimension of D. In Table 4.1, more or less optimal parameters are given, together with the value of $\tilde{\sigma}$, denoting the spectral radius of $\partial \mathbf{F}/\partial \mathbf{u}$, i.e.,

$$\tilde{\sigma} := \text{maximum of } |\alpha(\mu)| \text{ on the interval of eigenvalues } \mu \text{ of D.}$$

These parameters are determined on the basis of the model problem (2.1), where $d = -1/8$, $c = -1/2$, $\mathbb{R} := \{0, 1, \ldots, q\}$, and $q = 1$ until 8. These values were produced by the NAG routine E04 JAF for a suitable initial guess which was obtained by trial and error.

Although the above analysis is of restricted value (because of the deficiency in the frozen coefficient approach) and consequently the given parameters γ_r and ϖ_r will not be optimal, the values of $\tilde{\sigma}$ given in Table 4.1 are impressive small. In addition, our numerical experiments show that these parameters considerably increase the rate of convergence. We emphasize, however, that the given parameter sets are not unique with respect to minimizing the maximum of $|\alpha(\mu)|$. We found different sets of parameters which resulted in more or less the same damping factor $\tilde{\sigma}$.

Table 4.1. Smoothed SSOR parameters for the model problem.

r	γ_r	ϖ_r	γ_r	ϖ_r	γ_r	ϖ_r	γ_r	ϖ_r
0	1.29483	1.16429	1.52664	1.31372	1.43313	1.26135	1.39671	1.23887
1	0.67942	0.93100	0.83590	0.39225	0.59135	0.73279	0.61539	0.66521
2			0.49908	0.49426	0.23099	1.30867	0.25340	1.35313
3					0.19771	0.88092	0.19440	0.95840
4							0.18481	0.84568
	$q = 1,\ \tilde{\sigma} = .0066$		$q = 2,\ \tilde{\sigma} = .0028$		$q = 3,\ \tilde{\sigma} = .0043$		$q = 4,\ \tilde{\sigma} = .0043$	
0	1.92161	1.46464	1.45986	1.26980	1.29871	1.15502	1.27984	1.12992
1	0.90909	1.55156	0.76625	1.54084	0.80108	1.55903	0.45913	1.29275
2	0.24421	1.31166	0.12622	1.45593	0.24561	1.68935	0.31452	1.68786
3	0.73262	0.73607	0.18459	1.24836	0.16416	1.88198	0.25115	1.06478
4	0.32470	0.57986	0.09664	1.43038	0.21041	1.10603	0.38700	1.22488
5	1.24289	0.18750	0.07353	1.29864	0.20718	1.38827	0.10989	1.09238
6			0.19498	0.78077	0.09785	1.11741	0.21708	0.79981
7					0.11317	1.04884	0.08668	1.21313
8							0.09560	1.15921
	$q = 5,\ \tilde{\sigma} = .0041$		$q = 6,\ \tilde{\sigma} = .019$		$q = 7,\ \tilde{\sigma} = .025$		$q = 8,\ \tilde{\sigma} = .019$	

We conclude our analysis of smoothed SSOR with a picture of the behaviour of the function $\alpha(\mu)$. In Figure 4.1 this function is plotted for q=3.

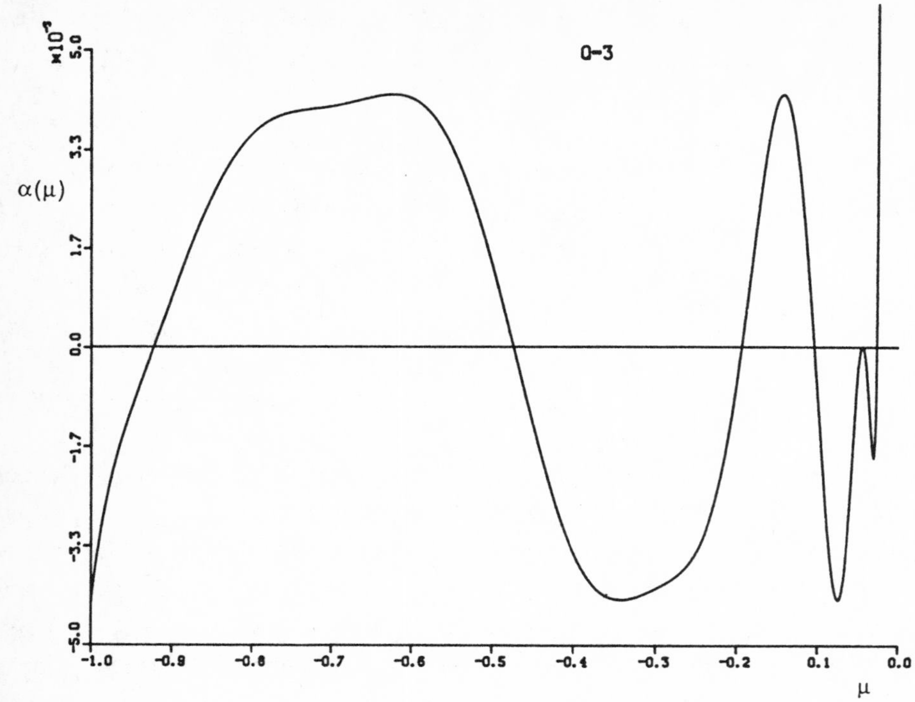

Figure 4.1. Behaviour of the function $\alpha(\mu)$ for $q = 3$.

4.2. Numerical experiments

By using residue smoothing employing the parameter values of Table 4.1, we obtain for the model problem the results as listed in Table 4.2. If, in addition, the Chebyshev acceleration is applied to this smoothed process, then the gain is almost negligible. This behaviour is similar to that observed in the preceding section on smoothed Jacobi iteration.

Table 4.2. Smoothed SSOR method for the model problem.

Chebyshev acceleration	[a,b]	Δx=1/16	Δx=1/64	Δx=1/256
no	--	**17**	39	64
yes	$[-\tilde{b},+\tilde{b}]$	18	**38**	**56**
yes	[-.01,+.01]	21	39	63
yes	[-.1,+.1]	27	50	74

The conclusion must be that the accelerating effect of the smoothing technique is so strong that the Chebyshev acceleration (which was invented to speed up traditional, unsmoothed basic iteration methods) is of no use in this case.

Next, we apply the same methods to the nonmodel problem (2.3). The analoque of Table 3.4 is given by Table 4.3. The conclusions that can be drawn from this table are the same as mentioned above. Finally, when compared with smoothed Jacobi, smoothed SSOR, provided with more or less optimal parameters, is slightly faster; however, the price to be paid is a more complicated algorithm.

Table 4.3. Smoothed SSOR method for problem (2.3).

Chebyshev acceleration	[a,b]	Δx=1/16	Δx=1/64	Δx=1/256
no	--	26	50	83
yes	$[-\tilde{b},+\tilde{b}]$	**26**	50	**83**
yes	[-.01,+.01]	26	50	83
yes	[-.1,+.1]	31	**49**	91
yes	[-.2,+.2]	37	58	110

REFERENCES

[1] HAGEMAN, L.A. & YOUNG, D.M. (1981), Applied iterative methods. Academic Press, New York.

[2] HOUWEN, P.J. VAN DER, BOON, C. & WUBS, F.W. (1988), Analysis

of smoothing matrices for the preconditioning of elliptic difference equations. To appear in Z. Angew. Math. Mech. 68.

[3] HOUWEN, P.J. VAN DER & SOMMEIJER, B.P. (1988), A fast algorithm for solving nonlinear elliptic difference equations on vector computers. In preparation.

[4] LYUSTERNIK, L.A. (1947), Trudy Mat. Inst. Steklov 20, pp. 49-64 (Russian).

[5] THIJE BOONKKAMP, J.H.M. ten (1987), Residual smoothing for accelerating the ADI iteration method for elliptic difference equations. Report NM-R8721, CWI, Amsterdam.

A NOTE ON PICARD-LINDELÖF ITERATION

OLAVI NEVANLINNA
Helsinki University of Technology
Institute of Mathematics
02150 Espoo, Finland

1. INTRODUCTION.

Iterative methods for initial value problems are well suited for parallel computation. A typical approach is to divide the interval where the solution is asked for, into subintervals, "windows", and to iterate inside the window before moving over to the next window. The following difficulty can arise. For stiff problems the transient phase is typically over at the end of the window so that the new starting value for the next window sits already on the "manifold" of smooth solutions. Usually one takes the initial function to be identically the initial value and it is no longer clear whether the iterates will stay smooth. A simple minded application of Picard-Lindelöf iteration (or "waveform relaxation") might then spend a lot of time in the early iterations since local error estimator would force an all too small step compared with the smoothness of the limit function. In [6] I have discussed the possibility of forcing the integration of the early sweeps to be done with larger steps than the important last sweeps.

In this note I want to discuss shortly the smoothness of the iterates. In terms of the kernel of the iteration operator one can give (in a model situation) a simple and intuitive necessary and sufficient condition to guarantee that the iterates inherite the smoothness of the limit function.

2. RESULTS.

Consider a system of initial value problem in the form

$$\dot{x} + Ax = f, \quad x(0) = x_0. \tag{2.1}$$

Here A is a constant $d \times d$ matrix and f can depend on time. We shall assume that the solution is being created by an iterate process, "Picard-Lindelöf iteration", as follows:

$$\dot{x}^{j+1} + Mx^{j+1} = Nx^j + f, \quad x^{j+1}(0) = x_0 \equiv x^0. \tag{2.2}$$

and each iterate x^j is computed from (2.2) approximately using a numerical integrator. In order to keep the discussion simple we idealize the integration to be done exactly, but measure the cost of integration as if the code would be based on first order local error estimation: a time step h is chosen to satisfy

$$h \,|\, \ddot{x}^j \,| = \varepsilon_j. \tag{2.3}$$

Thus the relevant measure for the cost or for the total number of time points is proportional to $\sum 1/\varepsilon_j \int | \ddot{x}^j |$. An efficient implementation of Picard-Lindelöf iteration would gradually decrease the tolerance ε_j. Here we shall not discuss the choice of ε but focus on estimating $\int | \ddot{x}^j |$.

We assume for simplicity that all eigenvalues of A snd M have positive real parts, some of them may be very large, and that there exists an initial value x_0 such that the corresponding solution of (2.1) is smooth. Further, the smooth solution is assumed to be represented as

$$x(t) = \sum t^i x_i \tag{2.4}$$

and - since we are interested in the second derivates - measure the smoothness on the window [0, 1] by

$$\| x \| := | x_0 | + | x_1 | + \sum_{i=2}^{\infty} i(i-1) \, | x_i |. \tag{2.5}$$

Taking the difference between (2.1) and (2.2) yields for $z^j := x - x^j$

$$\dot{z}^{j+1} + Mz^{j+1} = Nz^j, \quad z^j(0) = 0. \tag{2.6}$$

Introducing

$$k(t) := e^{-tM} N$$

and setting

$$k^{j*} = k * k^{(j-1)*}$$

we have

$$z^j = k^{j*} * (x - x_0). \qquad (2.7)$$

Substituting (2.4) into (2.7) yields

$$z^j(t) = \sum_{i=1}^{\infty} \int_0^t (t - s)^i \, k^{j*}(s) \; x_i \, ds \qquad (2.8)$$

and hence

$$\ddot{z}^j(t) = k^{j*}(t)x_1 + \sum_{i=2}^{\infty} i(i - 1) \int_0^t (t - s)^{i-2} \, k^{j*}(s) \; x_i \, ds. \qquad (2.9)$$

In order to estimate this we introduce the following bound:

$$B := \sup_j \; \sup_{|a|=1} \int_0^1 | \, k^{j*}(s)a \, | \, ds. \qquad (2.10)$$

We have immediately

Proposition 1.

$$\int_0^1 | \, \ddot{x} - \ddot{x}^j \, | \leq B \, \| \, x - x_0 \, \|. \qquad (2.11)$$

This result is sharp in the following way:

Proposition 2. For any given splitting M, N there exists $x_0 \neq 0$ and f such that $x(t) = (1 + t)\, x_0$, $\ddot{x} \equiv 0$ and for some j

$$\int_0^1 | \, \ddot{x}^j \, | = B \, \| \, x - x_0 \, \|. \qquad (2.12)$$

Proof. Since $\int |k^{j*}|$ tends to zero and a in (2.10) runs over a compact set there exists j and x_0 of unit length such that

$$B = \int_0^1 | k^{j*}(s)x_0 | \, ds.$$

If $f(t) = x_0 + (1 + t) Ax_0$ then $x(t) = (1 + t)x_0$ is the solution of (2.1). Since $x_1 = x_0$ and $\ddot{z}^j = - \ddot{x}^j$ we obtain from (2.9)

$$\int_0^1 | \ddot{x}^j | = B = B \, \| x - x_0 \|. \qquad \blacksquare$$

Combining Proposition 1 and 2 we conclude that - in this model situation - the important quantity $\int | \ddot{x}^j |$ stays small for smooth limit functions x iff B is of moderate size.

Example 1. For a scalar $\lambda \gg 1$ consider the splitting $0 < m = \lambda + n$ which gives $k(t) = e^{-tm} n$. The iteration (2.2) converges, in a pratical sense with rate ρ^j where $\rho = \frac{|n|}{m}$. In fact

$$\int_0^1 | k^{j*} | = \left(\frac{|n|}{m}\right)^j \frac{\Gamma_m(j)}{\Gamma(j)}$$

where the incomplete Γ-function satisfies

$$\Gamma_m(j) = \int_0^m e^{-\tau} \tau^{j-1} d\tau \to \Gamma(j) \quad \text{as} \quad m \to \infty.$$

In particular, as soon as the splitting yields a convergent iteration, B < 1 and so B is of "moderate size".

Remark 1. Even if $\int | \ddot{x}^j |$ is of moderate size the step sizes can locally be very small. Set, for example, in Example 1 $f(t) = 1 + \lambda(1 + t)$, $x_0 = 1$. Then $| \ddot{x}^j | = | k^{j*} |$ and by Stirling's formula

$$|\ddot{x}^j|_\infty \approx \frac{|n|\rho^{i-1}}{\sqrt{2\pi j}}$$

even so $\int |\ddot{x}^j| \approx \rho^j$. Thus the smallest steps are on the level of $\frac{\varepsilon^j}{|n|}$.

In order to obtain a situation where the process would converge but B would be large we have to look at higher dimensions.

Example 2. Let M = mI and N = mI - A. Then

$$k^{j*}(t) = \frac{e^{-mt}\, t^{j-1}}{(j-1)!} N^j$$

and

$$B = \sup_j \int_0^1 |k^{j*}| = \sup_j \frac{|N^j|}{m^j} \frac{\Gamma_m(j)}{\Gamma(j)} .$$

For large m we have convergence in practise if $\frac{\rho(N)}{m} < 1$. However B can be very large if initially $|N^j| \gg \rho(N)^j$. For example, for

$$N = \begin{pmatrix} 0 & n \\ 0 & 0 \end{pmatrix}$$

the process converges in two sweeps, no matter what n is, while for $|n| \gg m$ B is very large.

For homogeneous equations we have smooth solutions exactly when the initial value is in the eigenspace corresponding to the small eigenvalues of A. In that case the essential requirement is that M and N are invariant in that subspace. In particular, for M = mI this always follows. The previous example shows that for nonhomogeneous equations this however is not sufficient.

We end this with a special case where the "Jacobi-splitting" yields both convergence and a moderate B.

Let us call A *essentially strictly diagonally dominant* if there exist a moderate $\gamma > 0$ and some $\beta < 1$ such that for all i

$$\beta(a_{ii} + \gamma) \geq \sum_{j \neq i} |a_{ij}|.$$

If we use tha max-norm in space and the sup-norm in time it is easy to show that

$$| x - x^j |_\infty \leq e^\gamma \beta^j | x - x_0 |_\infty.$$

In the same way one shows

Proposition 3. Let M be the diagonal of A, the norm as above, then for each essentially strictly diagonally dominant A we have

$$B \leq e^\gamma.$$

I enclose a reference list on recent related work.

REFERENCES

[1] C.W.GEAR. The Potential for Parallelism in Ordinary Differential Equations. Report No. UIUCDCS-R-86-1246, Dept. of Computer Science, Univ. of Illinois at Urbana-Champaign, February 1986.

[2] E.LELARASMEE, A.E.RUEHLI, A.L.SANGIOVANNI-VINCENTELLI. The Waveform Relaxation Method for Time-Domain Analysis of Large Scale Integrated Circuits. IEEE Trans. Computer-Aided Design of ICAS, vol. CAD-1, no. 3, pp. 131-145, 1982.

[3] U.MIEKKALA. Dynamic Iteration Methods Applied to Linear DAE Systems. REPORT-MAT-A252, Helsinki University of Technology, Insitute of Mathematics, November 1987.

[4] U.MIEKKALA, O.NEVANLINNA. Convergence of Dynamic Iteration Methods for Initial Value Problems. SIAM J. Sci. Stat. Comp., Vol. 8, No. 4, 1987.

[5] U.MIEKKALA, O.NEVANLINNA. Sets of Convergence and Stability Regions. BIT 27 (1987), 554-584.

[6] O.NEVANLINNA. Remarks on Picard-Lindelöf iteration. REPORT-MAT-A254, Helsinki University of Technology, Institute of Mathematics, December 1987.

[7] R.D.SKEEL. Waveform Iteration and the Shifted Picard Splitting. CSRD Rpt. No. 700, Center for Supercomputing Research & Development, Univ. of Illinois at Urbana-Champaign, November, 1987.

ASPECTS OF PARALLEL RUNGE-KUTTA METHODS

SYVERT P. NØRSETT - HARALD H. SIMONSEN
Division of Mathematical Sciences,
Norwegian Insitute of Technology
Trondheim-Norway

Abstract. So far ODE-solvers have been implemented mostly on sequential computers. This has lead to development of methods that are very difficult to parallelisize. In this paper we discuss how to develop Runge-Kutta methods that lead to parallel implementation on computers with a small number of CPU's. Both explicit and implicit methods are discussed. Some initial experiments on a 2 processor CRAY X-MP and a 6 processor Alliant are presented.

1. INTRODUCTION.

Runge-Kutta methods (RK-methods) were initially developed for sequential computation. Indeed in the classical explicit methods, the computation must proceed in a sequential manner. For the ODE-system

$$y'(x) = f(y), \qquad x \geq a \tag{1}$$

$$y(a) = y_0, \qquad y_0 \in \mathbf{R}^s,$$

classical RK-methods take the form

$$k_i = f(y_n + h \sum_{j=1}^{i-1} a_{ij}\, k_j), \qquad i = 1, \dots , m \tag{2}$$

$$y_{n+1} = y_n + h \sum_{i=1}^{m} b_i\, k_i,$$

where h is the stepsize. Here k_i depends on k_i, ... k_{i-1} only.

There seems to be no room for parallel computation in these methods. However, by using directed graphs to represent the Runge-Kutta matrix, the results of Iserles and Nørsett [6] give us some insights into the capacities for parallelism in explicit Runge-Kutta methods. This is discussed in section 3.

Fully implicit RK-methods appear when the upper summation index in (2) is replaced by m.

$$k_i = f(y_n + h \sum_{j=1}^{m} a_{ij} k_j), \quad i = 1, \ldots m \tag{3}$$

All the k values, also called the stages of the RK-method, depend on each other. Using vector-notation and predictor-corrector iteration on implicit RK-methods, we can find methods that are of order up to 2m, but still retain some of the explicit flavour. Section 2 gives a summary of these results by Jackson and Nørsett [7].

For stiff systems, implicit methods are needed. Parallel versions of classical methods are described in section 4 while section 5 describes parallel collocation methods.

Finally in section 6 we show the results of some initial experiments with parallel RK-methods.

So far we have only mentioned parallelity accross the method. However it is also possible to use RK-methods to exploit parallelity accross the system, Skålin [11].

2. PREDICTOR-CORRECTOR METHODS.

Instead of writing implicit Runge-Kutta methods in terms of k-values we will use the stage-value notation,

$$Y_i = y_n + h \sum_{j=1}^{m} a_{ij} f(Y_j), \quad i = 1, \ldots, m$$

$$y_{n+1} = y_n + h \sum_{i=1}^{m} b_i f(Y_i).$$

Let us define the stage-vector Y by

$$Y = \left[Y_1^T, \ldots, Y_{m+1}^T\right]^T$$

and

$$f(Y) = \left[f(Y_1)^T, \ldots, f(Y_{m+1})^T\right]^T,$$

where $Y_{m+1} = y_{n+1}$ and $Y \in \mathbf{R}^{sm}$. The RK-matrix A is given by

$$A = \{a_{ij}\}_{i\ j=1}^{m}$$

and $\bar{A}$ by

$$\bar{A} = \begin{bmatrix} A & 0 \\ b^T & 0 \end{bmatrix}, \quad b = [b_1, \ldots, b_m]^T.$$

The RK-method can now be defined by

$$Y = e \otimes y_n + h\,\bar{A} \otimes f(Y) \tag{4}$$

$$y_{n+1} = Y_{m+1},$$

where $e = [1, \ldots, 1]^T \in \mathbf{R}^{m+1}$ and $\otimes$ is the tensor product.

It seems natural to use functional iteration to solve (4). Then

$$Y^{k+1} = e \otimes y_n + h\,\bar{A} \otimes f(Y^k), \quad k \geq 0 \tag{5}$$

where Y^0 is the initial guess for the exact solution Y. Let us remark that $Y^k \equiv Y$ for $k \geq m$ when A represents an explicit method. This is independent of Y^0.

A quick look at (5) shows that if Y^k- Y has order q, i.e. $Y^k - Y = 0(h^{q+1})$ then $Y^{k+1} - Y = 0(h^{\min(p,q+1)})$. Here p is the order of the method. A more precise result is:

Theorem 1. *Let* Y_e *be defined by*

$$Y_e = [y(x_n + c_i h); \quad i = 1, \ldots, m+1]^T \tag{6}$$

$$c_i = \sum_{j=1}^{m} a_{ij}, \quad c_{m+1} = 1. \tag{7}$$

Then if

$$Y^0 - Y_e = 0(h^{q+1})$$

we have

$$Y^{k}_{m+1} - y_{n+1} = 0(h^{\min(q+k,p)+1}).$$

The proof is given in Jackson and Nørsett [7].

Another interesting result is

Corollary 1. *If the weights* b_i *in* $\bar{A}$ *depend on a parameter* τ, $b_i = b_i(\tau)$, *and the order is* p *for all* $\tau \in \mathbf{R}$, *then the result in Theorem 1 is still valid.*

Let us remark that similar results can be obtained if the methods are written in derivative notation, $k_i = f(Y_i)$.

3. DIGRAPHS AND EXPLICIT RK-METHODS.

The RK-matrix A is visualized by the *digraph* G = (V, E). V = {1, ..., m} is the set of vertices whereas (i, j) belongs to the set of *directed edges* E if $1 \le i, j \le m$ and $a_{ij} \neq 0$. Different methods and digraphs are listed in figure 1.

In our plots od digraphs we break with convention: if $(i, j) \in E$ then the arrow in the line linking these vertices points at i. This has been done to make explicit the connection between the digraph and the flow of information in a parallel implementation. Similar effects could have been obtained by considering the digraph of A^T.

The digraph models the sparsity pattern of A. Figure 1 gives us four examples of digraphs which will be of interest later. Note that, as the sparsity pattern, rather than specific non-zero values of coefficients, matters, we use x and o to denote non-zero and zero values respectively.

We say that the vertices $v_1, ..., v_s$ where $s \ge 2$, form an *s-loop* if

a) $v_i \neq v_{i+1}$, i =1, ..., s - 1 and $v_1 \neq v_s$; and
b) either the pairs (v_i, v_{i+1}), i = 1, ..., s - 1, (v_s,v_1) of the pairs (v_{i+1}, v_i), i = 1, ..., s - 1, (v_1,v_s) all belong to E.

Note that the vertices are not assumed to be distinct. Moreover, the vertex v is a *1-loop* if $a_{v,v} \neq 0$, otherwise it is a *0-loop*.

Let σ be the length of the longest loop in the digraph of a Runge-Kutta matrix A. Clearly, it means that σ distinct stages are linked together in A into an implicit irreducible block and that no larger number of stages possesses this feature. Thus, we say that the underlying method is σ-implicit. In this terminology an explicit scheme corresponds to a 0-implicit Runge-Kutta method and a diagonally-implicit to a 1-implicit method.

Let us assume that the set V partitioned into ν disjoint non-empty subsets,

$$\{1, ..., m\} = \bigcup_{i=1}^{\nu} P_i, \quad P_1, ..., P_\nu \neq 0, \quad P_i \cap P_j = \emptyset \quad \forall i \neq j. \tag{8}$$

We call (8) a *processor partitioning* of the method. We further partition each P_l,

$$P_l = \bigcup_{i=1}^{\mu} L_{l,i}, \qquad l = 1, ..., \nu, \tag{9}$$

where $L_{l,i} \cap L_{l,j} = \emptyset$ for all $i \neq j$. Note that the $L_{l,i}$'s are allowed to be empty sets. However, we assume that for some $r \in \{1, ..., \nu\}$ it is true that $L_{r,1}, ..., L_{r,\mu} \neq \emptyset$.

We say that (9) is a *level partitioning* of the method if (a) for all $l = 1, ...,\nu$,$j =$ $= 1 ..., \mu$ all the vertices in $L_{l,j}$ belong to a loop; and (b) if $a_{ij} \neq 0$ then *either* there exists a $q \in \{1, ..., \nu\}$, $1 \leq r \leq s \leq \mu$ such that $i \in L_{q,s}$, $j \in L_{q,r}$, *or* there exists $q.t \in \{1, ..., \nu\}$, $q \neq t$, $1 \leq r \leq s \leq \mu$ such that $i \in L_{t,s}$, $j \in L_{q,r}$. A combination of processor and level partitioning of a σ-implicit Runge-Kutta method will be termed a $\{\nu, \mu, \sigma\}$*-implementation*. Figure 2 presents a $\{2, 2, 2\}$-implementation of the method III from figure 1.

Scheme	*Runge-Kutta matrix*	*Digraph*
I	$\begin{bmatrix} \times & \times & o & o \\ \times & \times & o & o \\ o & o & \times & \times \\ o & o & \times & \times \end{bmatrix}$	
II	$\begin{bmatrix} \times & o & o & o \\ o & \times & o & o \\ \times & \times & \times & o \\ \times & \times & o & \times \end{bmatrix}$	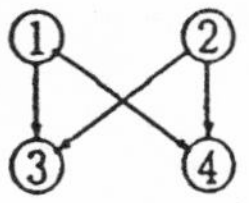
III	$\begin{bmatrix} \times & \times & o & o & o & o \\ \times & \times & o & o & o & o \\ o & o & \times & \times & o & o \\ o & o & \times & \times & o & o \\ \times & \times & \times & \times & \times & o \\ \times & \times & \times & \times & o & \times \end{bmatrix}$	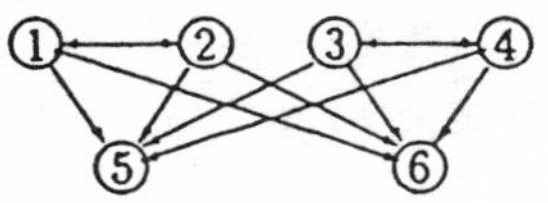
IV	$\begin{bmatrix} \times & o & o & o & o & o \\ o & \times & o & o & o & o \\ o & o & \times & o & o & o \\ \times & \times & \times & \times & o & o \\ \times & \times & \times & o & \times & o \\ \times & \times & \times & o & o & \times \end{bmatrix}$	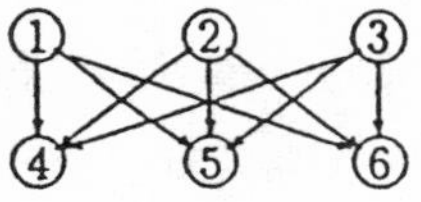

Fig. 1 : Runge-Kutta matrices and digraphs.

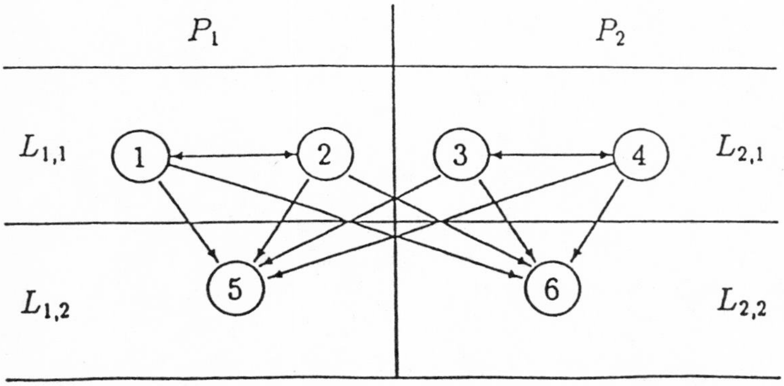

Fig. 2 : A {2,2,2}-implementation of the method III from fig. 1.

The intuitive meaning of level partitioning is clear: Vertices in each $L_{l,i}$ must "communicate" with all the remaining vertices in that set. They are disconnected from vertices in all $L_{k,i}$, $k \neq l$ and they are allowed only to "transmit" to, but not to "receive" from, vertices in the sets $L_{k,j}$, $k > l$.

Using this language we can prove that:

Theorem 2. *The order* p *of an explicit Runge-Kutta method with a* $\{\nu, \mu, \sigma\}$*-implementation cannot exceed* μ*, regardless of the number of processors* ν.

The proof can be found in Iserles and Nørsett [6].

This means that using blocks of explicit methods in the RK-matrix can not helps us to increase the order of the overall method.

It is well known that for explicit methods the order p is bounded by the number of stages. When the number of stages is less than 4, the order can equal the number of stages. For stage numbers greater than 4 the order of an explicit Runge-Kutta method is always lower than the number of stages. Our hope is to be able to get order p using p parallel-stages (p-stages). This is the case for the following scheme by Butcher,

$$
\begin{array}{c|cccccc}
0 & & & & & & \\
\frac{1}{8} & \frac{1}{8} & & & & & \\
\frac{1}{4} & 0 & \frac{1}{4} & & & & \\
\frac{1}{2} & -\frac{1}{2} & 1 & 0 & & & \\
\frac{3}{4} & \frac{21}{16} & -\frac{15}{8} & \frac{3}{4} & \frac{9}{16} & & \\
1 & -\frac{17}{7} & 4 & 0 & -\frac{12}{7} & \frac{8}{7} & \\
\hline
 & \frac{7}{90} & 0 & \frac{32}{90} & \frac{12}{90} & \frac{32}{90} & \frac{7}{90}
\end{array}
$$

It is straight forward to show that this scheme admits the level partitioning

$$
\begin{array}{ll}
L_{1,1} = \{1\} & L_{2,1} = \emptyset; \\
L_{1,2} = \{2\} & L_{2,2} = \{3\}; \\
L_{1,3} = \{4\} & L_{2,3} = \emptyset; \\
L_{1,4} = \{5\} & L_{2,4} = \emptyset; \\
L_{1,5} = \{6\} & L_{2,5} = \emptyset.
\end{array}
$$

This is a $\{2, 5, 0\}$-implementation and we can get order 5 using 5 p-stages.

An interesting question arises: *Is it always possible to find explicit Runge-Kutta methods of order* p *using* p *parallel stages?*

4. PARALLEL DIRK-METHODS.

Stiff equations require implicit methods. On sequential computers SDIRK or SIRK methods work very efficiently. The STRIDE code of Burrage, Butcher and Chipman [1] and the SIMPLE code of Nørsett and Thomsen [9] are all well known codes based on these methods.

For parallel processing we would like to find methods that have high order and a small number of parallel stages. A very simple idea is to have methods of the form

$$
A = \begin{bmatrix} c_1 & \cdots & 0 \\ \cdots & \cdots & \cdots \\ \cdots & \cdots & \cdots \\ 0 & \cdots & c_m \end{bmatrix}.
$$

In Jackson and Nørsett [7] methods of this type are analyzed. Unfortunately the maximum attainable order for these methods is 2.

Iserles and Nørsett [6] discuss block methods of the form

$$
A = \begin{bmatrix} A_1 & \cdots & 0 \\ \cdots & \cdots & \cdots \\ \cdots & \cdots & \cdots \\ 0 & \cdots & A_\nu \end{bmatrix} \tag{10}
$$

or

$$A = \begin{bmatrix} \hat{A} & \cdots & \cdots & 0 \\ D_{21} & \cdots & & \cdots \\ \cdots & & \cdots & \cdots \\ D_{\mu 1} & \cdots & D_{\mu\mu-1} & \hat{A} \end{bmatrix}. \tag{11}$$

The first kind of method (10) is termed type I, while the second (11) is termed type II. A_i, $\hat{A}$ and D_{ij} are block matrices.

When the A_i's in the type I methods are $\mu\times\mu$-matrices, we write the method as follows:

$$\begin{array}{c|c} c_1 & A_1 \\ \hline \rho_1 & b_1^T \end{array} + \ldots + \begin{array}{c|c} c_\nu & A_\nu \\ \hline \rho_\nu & b_\nu^T \end{array}. \tag{12}$$

Using this notation, y_{n+1} can be written as

$$y_{n+1} = \rho_1 Y_{n+1}^{(1)} + \ldots + \rho_\nu Y_{n+1}^{(\nu)} \tag{13}$$

Here each processor computes μ stages.

An example of a {2, 2, 2}-implementation with internal order 3 is the following L-stable, 2 p-stage method:

$$\begin{array}{c|cc} \frac{11-\sqrt{31}}{18} & \frac{5}{12}+\frac{\sqrt{31}}{186} & \frac{7}{36}-\frac{17\sqrt{31}}{279} \\ \frac{11+\sqrt{31}}{18} & \frac{7}{36}+\frac{17\sqrt{31}}{279} & \frac{5}{12}-\frac{\sqrt{31}}{186} \\ \hline 3 & \frac{1}{2}+\frac{\sqrt{31}}{31} & \frac{1}{2}\frac{\sqrt{31}}{81} \end{array} + \begin{array}{c|cc} \frac{1}{3} & \frac{1}{2} & -\frac{1}{6} \\ 1 & \frac{1}{2} & \frac{1}{2} \\ \hline -2 & \frac{3}{4} & \frac{1}{4} \end{array}$$

The combined order is 4.

Type II methods seem to be more powerful. However they require more interprocessor communication. An example of such a method was given by Jackson ans Nørsett in [7]. This was a {2, 2, 1} method with order 4. Unfortunately it was not A-stable. Iserles and Nørsett [6] have further examined these kind of methods. The following method is of order 4 and is also A-stable.

$$
\begin{array}{c|cccc}
\frac{1}{2} & \frac{1}{2} & & & \\
1 & 0 & 1 & & \\
\frac{1}{2} & \frac{3}{2} & -\frac{3}{2} & \frac{1}{2} & \\
0 & -3 & 2 & 0 & 1 \\
\hline
 & \frac{1}{3} & \frac{1}{6} & \frac{1}{3} & \frac{1}{6}
\end{array}
$$

This method is also embedded in a third order method which can be used for local error control.

$$
\begin{array}{c|cccccc}
\frac{1}{2} & \frac{1}{2} & & & & & \\
1 & 0 & 1 & & & & \\
\frac{1}{2} & \frac{3}{2} & -\frac{3}{2} & \frac{1}{2} & & & \\
0 & -3 & 2 & 0 & 1 & & \\
\frac{3}{2} & \frac{1}{3} & \frac{1}{6} & \frac{1}{3} & \frac{1}{6} & \frac{1}{2} & \\
2 & \frac{1}{3} & \frac{1}{6} & \frac{1}{3} & \frac{1}{6} & 0 & 1 \\
\hline
 & -\frac{7}{6} & 1 & \frac{8}{9} & \frac{1}{2} & -\frac{1}{6} & -\frac{1}{18}
\end{array}
$$

5. PARALLEL COLLOCATION METHODS.

In STRIDE collocation methods are used. Laguerre functions are used as collocation polynomials. Here the eigenvalues of the Runge-Kutta matrix are all real and equal. This leads to efficient implementation on a sequential computer. These methods all have order m using m stages. Using m processors the idea is to let each processor do one function evaluation in addition to the linear algebra needed.

Using tensor notation we write Runge-Kutta methods on the form

$$R(Y) := Y - e \otimes y_n - h\,A \otimes f(Y) = 0, \qquad A \in \mathbf{R}^{m \times m} \tag{14}$$

Let us assume that A has real and distinct eigenvalues, $\gamma_1, ..., \gamma_m \in \mathbf{R}^+$. Then a nonsingular matrix T exists such that

$$A = T^{-1} DT, \qquad D = \mathit{diag}\ \{\gamma_1, ..., \gamma_m\}$$

For stiff systems we use modified Newton iteration

$$\begin{aligned} N\Delta^{k+1} &= - R(Y^k), \\ Y^{k+1} &= y^k + \Delta^{k+1}, \end{aligned} \tag{15}$$

where $k \geq 0$ and Y^0 is computed by extrapolating the local interpolant from the step before. N is defined by

$$\begin{aligned} N &= I \otimes I - hA \otimes J \\ &= T^{-1}\,(I \otimes I\ - hD \otimes\ J) \otimes T \end{aligned} \tag{16}$$

Here J is the Jacobian $\partial f/\partial y$. The iteration can then be transformed to

$$(I \otimes I - hD \otimes J)z = F \tag{17}$$

with

$$\begin{aligned} z &= T \otimes \Delta^{k+1} = \left[z_1^T, ..., z_m^T\right]^T \\ F &= - T \otimes R(Y^k) = \left[F_1^T, ..., F_m^T\right]^T. \end{aligned} \tag{18}$$

Since D is diagonal, we can split this into m systems

$$(I - h\gamma_i J)\ z_i = F_i, \qquad i = 1, ..., m. \tag{19}$$

Each of these systems is independent of the other. This means that we can solve all of them on different processors at the same time.

Methods with a one-point real spectrum can be constructed by collocation based on the zeros of the Laguerre polynomials. The fundamental question is then which zeros to use for collocation in order to obtain a Runge-Kutta method with a real, distinct spectrum.

The answer can be found by using the theory of bi-orthogonal polynomials, Iserles and Nørsett [5].

Let us define Q(z) by

$$Q(z) = \prod_{i=1}^{m} (1 - c_i z) = \sum_{i=0}^{m} S_{m-i}(c_1, ..., c_m)\, z^i \tag{20}$$

$$= z^m \sum_{i=0}^{m} S_i(c_1, ..., c_m) \frac{1}{z^i}$$

where S_0, ..., S_m are well known symmetric functions. Q(z) is the wanted denominator of the rational stability function for our collocation method. According to Nørsett and Wanner [10] the collocation polynomial M(t) for an m-point collocation method is then given by

$$M(t) = \sum_{i=0}^{m} \frac{1}{i!} S_i(c_1, ..., c_m)\, t^i. \tag{21}$$

In order to prove that M(t) has only distinct and real zeros, namely the collocation points, we rewrite Q(z) as

$$Q(z) = (-z)^m \prod_{i=1}^{m} c_i \sum_{k=0}^{m} k! a_k(c_1, ..., c_m) \frac{1}{z^k}. \tag{22}$$

Then, Iserles and Nørsett [4],

$$P_m(t) = \sum_{k=0}^{m} a_k(c_1, ..., c_m) t^k \tag{23}$$

is the bi-orthogonal polynomial based on the weight-function $w(x, \mu)$,

$$w(x, \mu) = e^{-\mu x}$$

and measure

$$\alpha(x) = x, \qquad x \in (0, \infty).$$

It is easily seen that

$$P_m(t) = \frac{(-1)^m}{\prod_{j=1}^{m} c_j} M(t).$$

Since $P_m(t)$ is a bi-orthogonal polynomial we can conclude that $P_m(t)$, and hence $M(t)$ has real, distinct and positive zeros. Further we have the following representation for $M(t)$,

$$M(t) = \prod_{i=1}^{m} \frac{T_{c_i} t^m}{m!} \tag{24}$$

where T_c is the operator

$$(T_c f)(x) = f(x) - cf'(x). \tag{25}$$

The following results can now be stated:

Theorem 3. *Collocation based on the collocation polynomial* $M(t)$ *in* (24) *produces a* SIRK *method with real and distinct eigenvalues* $c_1, ..., c_m$.

Corollary 2. *If* $c_i \to c$, *for* $i = 1, ... c_m$, *these* SIRK-*methods reduce to the* SIRK-*methods in* [2].

It is still an open question how to choose the $c_1, ..., c_m$ values in order to get a stable method with a minimal error.

6. NUMERICAL EXPERIMENTS.

Some initial experiments have been done on a 2 processor CRAY X-MP and on a 6 processor Alliant. More experiments will be done in the near future.

From Lie [8] we have the following examples. The Hammer-Hollingsworth method

$$\begin{array}{c|cc} \frac{1}{2}-\frac{\sqrt{3}}{6} & \frac{1}{4} & \frac{1}{4}-\frac{\sqrt{3}}{6} \\ \frac{1}{2}+\frac{\sqrt{3}}{6} & \frac{1}{4}+\frac{\sqrt{3}}{6} & \frac{1}{4} \\ \hline & \frac{1}{2} & \frac{1}{2} \end{array}$$

was implemented as a predictor-corrector method of the type described in section 2. Y^0 is computed by a Hermite interpolation polynomial based on the y- and y'-values from the last step. From the result of section 2 we know that one iteration should suffice to get the order of the basic method. The implementation were done on a CRAY X-MP/28 using miscrotasking. As a test problem, van der Pol's equation with $\varepsilon = 5$ was used. Speedup factors are given in table 1. The code uses 1-iteration on the average. With the high

Integration interval	Speedup
[0,20]	1.82
[0,50]	1.90
[0,100]	1.90

Table 1 : Speedup factors on the CRAY X-MP/28

parallelism shown, the H-H method will effectively use 2 parallel function evaluations per step, compared to 4 for a small explicit RK-method.

On an Alliant FX/8 some experiments using two 3-stage collocation based methods were done. Predictor-corrector iteration was used. The methods implemeted were a 3-stage method of order 6 by Butcher and the 3 stage Radau IIA method. The collocation polynomial was used to compute Y^0. A modified version of the restricted 3-body problem was the test equation. The results for 3 and 6 processors are given in table 2.

The maximum attainable speedup is not 3 because there is some inherent sequential work. The experiments were performed with no specific optimization, so the results presented seem resonable.

No of processors	Butcher order 6	Radau IIA
3	2.52	2.61
6	2.52	2.61

Table 2 : Speedup factors on the Alliant FX/6.

REFERENCES

[1] K.BURRAGE, J.C.BUTCHER and F.H.CHIPMAN, An implementation of singly implicit Runge-Kutta methods. BIT 20 (1980), pp. 326-340.

[2] J.C.BUTCHER, A transformed implicit Runge-Kutta method. J.Ass. Comput. Mach., Vol. 26 (1979), pp. 731,738.

[3] E.HAIRER, S.P.NØRSETT, G.WANNER, Solving Ordinary Differential Equations I : Nonstiff Problems. Springer-Verlag, 1987.

[4] A.ISERLES and S.P.NØRSETT, Bi-orthogonal polynomials. Lecture notes in mathematics. Vol. 1171 (1984), Springer-Verlag, pp. 92-100.

[5] A.ISERLES and S.P.NØRSETT, On the theory of Bi-orthogonal polynomials. Trans. of Amer. Math. Soc. 1988, pp. 455-474.

[6] A.ISERLES and S.P.NØRSETT, On the theory of parallel Runge-Kutta methods. To appear, 1988.

[7] K.JACKSON and P.J.NØRSETT, Parallel Runge-Kutta methods. To appear 1988.

[8] I.LIE, Some aspects of parallel Runge-Kutta methods. Tech. rep. no. 3/87, Div. Math. Sci., Norwegian Inst. of Tech., Trondheim.

[9] S.P.NØRSETT and P.THOMSEN, SIMPLE-a stiff system solver. Tech. rep. 1987, Div. of Math. Sci., Norwegian Inst. of Tech., Trondheim.

[10] S.P.NØRSETT and G.WANNER, The real-pool sandwich for rational approximations and oscillation equations. BIT 19 (1979), pp. 79-84.

[11] R. SKÅLIN, Numerisk løsning av ordinœre differensialligninger ved parallellisering påsystemnivå. Et studium av Runge-Kutta metoder. Master Thesis 1988 (In norwegian).

TOLERANCE PROPORTIONALITY IN ODE CODES

L.F.SHAMPINE(*)
Mathematics Department
Southern Methodist University
Dallas, Texas 75275
U.S.A.

Abstract. Some bounds on the global error of the numerical solution of the initial value problem for a system of ordinary differential equations (ODEs) are developed that show the behavior with respect to the local error tolerance. Problems with solutions that are not smooth at a few points are considered. It is shown how to do local extrapolation with the backward differentiation formulas. An error criterion is proposed that has most of the advantages of both error per step and error per unit step.

1. INTRODUCTION.

We are concerned with the numerical solution of the initial value problem for a system of ordinary differential equations (ODEs),

(1) $$y' = f(x, y) \quad , \qquad a \leq x \leq b \quad ,$$

(2) $$y(a) = A \quad .$$

Codes for this purpose step from a to b producing approximations to y(x) on a mesh. The mesh points are chosen automatically so that the error made at each step is less than a tolerance specified by the user. Controlling this local error provides only an indirect control of the global error, the difference between the approximate solution and the true solution y(x). For this reason it is very common that users solve a problem (1,2) with several tolerances and assess the global errors by comparing the results.

(*) This work was partially supported by the Applied Mathematical Sciences program of the Office of Energy Research under DOE grant DE-FG05-86ER25024.

There are a number of criteria for controlling the local error. One, EPUS, is well-known to provide a bound on the global error that it proportional to the tolerance. After some preliminaries in Section 2, we give a sharper version of this bound in Section 3. Another criterion, EPS, is much more popular than EPUS. As we show by example, a bound like that for EPUS is not true for this criterion. When this fact was first realized it bothered greatly people trying to develop the theory of ODE solvers, but it did not stop the people writing codes from using EPS. In [6] we explained what is going on. The essence of the matter is that in practice an *efficient* set of mesh points is chosen. Here we make a realistic assumption to this effect and then prove a bound like that for EPUS. There is a variation in the way an integration is advanced called local extrapolation. It was observed in [9] that EPS with local extrapolation, XEPS, is a kind of generalized EPUS. In Section 3 we go into this a little to see how it combines the advantages of EPS and EPUS.

The theory of ODE solvers supposes that f is smooth. Quite a lot of experimental evidence says that the popular codes can integrate problems that have a few points x where the solution is not smooth. In Section 4 we investigate a class of such problems when solved by explicit Runge-Kutta methods. It is possible to explain the experimental results using the analysis of Section 3.

It is generally thought that backward differentiation formula (BDF) codes cannot explain local extrapolation. In Section 5 we explain how to do this. Local extrapolation is not natural to formulas like the BDF. In Section 6 we present a new error criterion that also enjoys many of the advantages of EPS for a given tolerance along with the advantage of EPUS when the tolerance is changed.

Computations are reported in the last section that illustrate some of the theoretical developments of the paper.

2. PRELIMINARIES.

The task is to solve numerically the initial value problem for a system of ordinary differential equations,

(1) $$y' = f(x, y) \quad , \qquad a \le x \le b \quad ,$$

(2) $$y(a) = A \quad .$$

Unless we state the contrary, f is assumed to be as smooth as necessary for the arguments made.

The methods considered start with the given solution value at $x_0 = a$ and step to b producing approximate solution vectors $y_n \approx y(x_n)$ at $a = x_0 < x_1 < x_2 < \ldots < x_N = b$. The step size h_n is defined by

$$h_n = x_{n+1} - x_n.$$

At each step the code is to control the local error made. This quantity is defined at x_n in terms of the solution u(x) of

(3) $$u' = f(x, u) \quad , \qquad u(x_n) = y_n \quad .$$

The local error in stepping from x_n is

$$le_n = u(x_n + h_n) - y_{n+1}.$$

It is supposed that the method is of order p and has the principal error function $\phi(x, y)$ so that

$$le_n = h_n^{p+1} \phi(x_n, y_n) + 0(h_n^{p+2}).$$

In the case of one-step methods it is usual to estimate the local error by taking each step with two formulas. The error of the result y_{n+1} of order p is estimated by comparing it to the result y^*_{n+1} of order (at least) p + 1 by

$$est = y^*_{n+1} - y_{n+1} \quad .$$

This is justified by

$$le_n = u(x_{n+1}) - y_{n+1} = y^*_{n+1} - y_{n+1} + (u(x_{n+1}) - y^*_{n+1}) = est + 0(h_n^{p+2}) \quad .$$

Whatever the method of estimating the local error, if

$$est = le_n + 0(h_n^{p+2}) \quad ,$$

it is obviously the case that the quantity y^*_{n+1} *defined* by

$$y^*_{n+1} = y_{n+1} + est$$

represents a method of order p + 1.

A norm $\| \cdot \|$ and a tolerance τ are specified by the user of the code. Two kinds of error control are seen. The criterion of error per unit step, EPUS, accepts a result y_{n+1} when

(4) $\| le_n \| \leq h_n \tau$.

The criterion of error per step, EPS, accepts the result when

(5) $\| le_n \| \leq \tau$.

We have observed that the error estimation procedures correspond to computing an approximate solution y^*_{n+1} of higher order. It is appealing to advance the integration with the result y^*_{n+1} instead of y_{n+1}. This is called local extrapolation. Combined with the two types of error control, EPUS and EPS, this yields two additional modes of computation, XEPUS and XEPS. All four ways of proceeding have been used in production codes.

3. BOUNDS OF THE GLOBAL ERROR.

The global, or true, error of an approximation y_n to the solution $y(x_n)$ of (1,2) is $y(x_n) - y_n$. There is a well-known result to the effect that if f satisfies a Lipschitz condition with respect to y, then control of the local error at each step by the criterion of EPUS implies a control of the global error. We give a sharper form of this result. We further show that a similar result is valid when a fixed number of steps is accepted by the criterion of EPS. Later we apply this result to the integration of an f that is not smooth. If EPS is used for all steps, it is not in general true that the global error is bounded independent of the number of steps. We clarify the matter and show how to obtain a bound with an additional, realistic assumption.

We suppose that the differential equation (1) is stable in the sense that there is a constant ρ such that for any two solutions v(x), w(x) of (1) and any x, Δ such that $a \leq x < x + \Delta \leq b$, we have

(6) $\| v(x + \Delta) - w(x + \Delta) \| \leq e^{\rho\Delta} \| v(x) - w(x) \|$.

The usual approach is based on the fact that if f satisfies a Lipschitz condition with constant L, this bound holds with $\rho = L$. In our sharper version we can have $\rho < L$, and even $\rho \leq 0$.

The idea of the proof is to relate the global error at x_{n+1} to the amplification allowed by (6) of the global error present at x_n and to the local error made in the step from x_n. We start with the identity

$$y(x_{n+1}) - y_{n+1} = u(x_{n+1}) - y_{n+1} + y(x_{n+1}) - u(x_{n+1})$$

where u(x) is the local solution (3). Taking norms and using (6) we are led to

$$\begin{aligned} \| y(x_{n+1}) - y_{n+1} \| &\le \| le_n \| + \| y(x_{n+1}) - u(x_{n+1}) \| \\ &\le \| le_n \| + e^{\rho h_n} \| y(x_n) - y_n \| \quad . \end{aligned}$$

Suppose that the code controls the local error according to EPUS, i.e., inequality (4) holds. It is now an easy induction to establish that for all n

$$(7) \quad \begin{aligned} &\| y(x_n) - y_n \| \le [\, \| y(x_o) - y_o \| + (x_n - x_o)\,\tau]\, e^{\rho(x_n - x_o)} && \text{if } \rho > 0 \\ &\| y(x_n) - y_n \| \le \| y(x_o) - y_o \| + (x_n - x_o)\,\tau && \text{if } \rho \le 0. \end{aligned}$$

In practice we always start with $y_o = y(x_o)$ so that

$$(8) \quad \begin{aligned} &\| y(x_n) - y_n \| \le \tau\, (x_n - x_o)\, e^{\rho(x_n - x_o)} && \text{if } \rho > 0, \\ &\| y(x_n) - y_n \| \le \tau\, (x_n - x_o) && \text{if } \rho \le 0. \end{aligned}$$

This holds for all $x_n \in [a, b]$ so that

$$\begin{aligned} &\max_n \; \| y(x_n) - y_n \| \le \tau\, (b - a)\, e^{\rho(b - a)} && \text{if } \rho > 0, \\ &\max_n \; \| y(x_n) - y_n \| \le \tau\, (b - a) && \text{if } \rho \le 0. \end{aligned}$$

Notice that this bound is independent of the number of steps. Indeed, the step size appears only in the requirement that it be small enough to yield the desired local error at each step. A key point is that the bound is proportional to the tolerance τ.

Suppose now that EPUS is used through the step to x_m, EPS is used for the step to x_{m+1}, and EPUS is used thereafter. Then

$$\begin{aligned} \| y(x_{m+1}) - y_{m+1} \| &\le \| le_m \| + e^{\rho h_m} \| y(x_m) - y_m \| \\ &\le \tau + e^{\rho h_m} \| y(x_m) - y_m \| . \end{aligned}$$

Let us consider the case $\rho > 0$, for which (8) shows that when $y_o = y(x_o)$,

$$\| y(x_{m+1}) - y_{m+1} \| \le \tau + \tau\, (x_m - x_o)\, e^{\rho(x_{m+1} - x_o)} \quad .$$

If we regard the integration as being restarted at x_{m+1}, we can apply (7) with y_{m+1} playing the role of y_0 and x_{m+1} the role of x_0. Thus for $n \geq m + 1$,

$$\| y(x_n) - y_n \| \leq [\, \| y(x_{m+1}) - y_{m+1} \| + \tau\,(x_n - x_{m+1})\,]\, e^{\rho(x_n - x_{m+1})} ,$$

which, when combined with the bound for the error of y_{m+1}, leads to

$$\| y(x_n) - y_n \| \leq \tau\, e^{\rho(x_n - x_{m+1})} + \tau\,(x_n - x_0)\, e^{\rho(x_n - x_0)}$$

for $\rho > 0$ and $n \geq m + 1$. A similar argument yields

$$\| y(x_n) - y_n \| \leq \tau\,(x_n - x_{m+1}) + \tau\,(x_n - x_0)$$

for $\rho \leq 0$ and $n \geq m + 1$.

These bounds show that a bounded number of steps taken with the criterion of EPS do not change the conclusion that a bound on the global error is proportional to the tolerance τ. Indeed, this is true even if we allow an error greater than τ, provided that it is $O(\tau)$ as $\tau \to 0$. Later we use the case $\| le_m \| \leq \delta\,\tau$ for a constant δ. The situation is quite different when all the steps are taken with EPS.

Consider the solution of

$$y' = 0 \;, \qquad 0 \leq x \leq 1 \;, \qquad y(0) = 1$$

by a method with the constant step size $h = 1/N$ as the integer $N \to \infty$. The method takes $y_0 = 1$, and for other n, $y_{n+1} = y_n + \tau$. It is easily verified that $le_n = \tau$. Now $y_N = 1 + N\tau$, so the global error

$$y_N - y(1) = N\tau = \frac{\tau}{h}$$

is not bounded independent of the number of steps N.

This example is not representative of computing practice because the step size chosen is not *efficient*. We did not need to concern ourselves with this matter for EPUS, but now we find that an additional hypothesis is necessary. To develop a suitable assumption, let us consider how the codes work. Basically they attempt to find an h_n such that (for EPS)

$$\| le_n \| \approx \beta\,\tau$$

for a constant β with $0 < \beta < 1$. Using the principal error function, this means that

$$h_n^{p+1} \| \phi(x_n, y_n) \| \approx \beta \tau$$

or

$$h_n \approx (\beta\tau / \| \phi(x_n, y_n) \|)^{1/(p+1)} .$$

(More details about step size selection are provided in [7].) We shall assume, then, that there is a constant $\gamma > 0$ such that

$$h_n \geq \gamma^{-1} \tau^{1/(p+1)}$$

for a method of order p. This amounts to supposing that however the step size is selected, it has at least the right order. Then

$$\| le_n \| \leq h_n \left(\frac{\tau}{h_n}\right) \leq h_n \left(\gamma \tau^{p/(p+1)} \right).$$

According to this we can modify the earlier result for EPUS with, say, $y_0 = y(x_0)$ by simply replacing τ there with $\gamma \tau^{p/(p+1)}$. This results in

$$\max_n \| y(x_n) - y_n \| \leq \tau^{p/(p+1)} \gamma (b - a) e^{\rho(b - a)} \qquad \text{if } \rho > 0$$

$$\max_n \| y(x_n) - y_n \| \leq \tau^{p/(p+1)} \gamma (b - a) \qquad \text{if } \rho \leq 0.$$

With the assumption that the step size chosen is reasonably efficient, we get the same kind of result that we got with EPUS, the essential difference being that the qualitative behavior with respect to a change of τ is altered. This is not an artifact of the method of proof. In [7] we modelled realistic codes with an asymptotic analysis and found the same behavior with respect to change of τ.

A popular procedure is XEPS. As pointed out in [9] this can be analyzed as a kind of generalized EPUS. Asymptotic approximations are derived in [7]. Here we argue more in the spirit of bounds. The step size is selected so that

$$\| u(x_{n+1}) - y_{n+1} \| \leq \tau ,$$

and we suppose that an efficient step size is selected. From the expression

$$u(x_{n+1}) - y_{n+1} = h_n^{p+1} \phi(x_n, y_n) + 0(h_n^{p+2})$$

we see that, with a natural non-degeneracy assumption, as $\tau \to 0$, so does h_n. Indeed, $h_n = 0(\tau^{1/(p+1)})$. The local error of the higher order result used to advance the integration is

$$le_n = u(x_{n+1}) - y^*_{n+1} = h_n^{p+2}\ \psi(x_n, y_n) + 0(h_n^{p+3}).$$

For all sufficiently small τ, we have

$$\| le_n \| \le \| u(x_{n+1}) - y_{n+1} \| \le \tau .$$

Our bound for EPS then applies to this method and we conclude that it converges as $\tau \to 0$. With convergence we can assert that

$$u(x_{n+1}) - y_{n+1} = h_n^{p+1}\ \phi(x_n, y(x_n)) + 0(h_n^{p+2})$$

and

$$le_n = h_n^{p+2}\ \psi(x_n, y(x_n)) + 0(h_n^{p+3}).$$

This implies that

$$\| le_n \| = h_n \left[\| u(x_{n+1}) - y_{n+1} \| \ \frac{\|\psi(x_n, y(x_n))\|}{\|\phi(x_n, y(x_n))\|} + 0(h_n^{p+2}) \right].$$

If we let

$$\nu = \max_{a \le x \le b} \frac{\|\psi(x,y(x))\|}{\|\phi(x,y(x))\|} ,$$

then

$$\| le_n \| \le h_n [\, \nu\, \tau + 0(\tau) \,] .$$

Thus XEPUS corresponds to an error per unit step with τ replaced by $\nu\,\tau + 0(\tau)$. In its leading term, the error bound is proportional to τ. There is a disagreeable unknown constant here, but it must be appreciated that we do not know the quantity ρ in the bound for EPUS either.

4. NON-SMOOTH F.

Although the typical problem has the f of (1) a smooth function, it is by no means unusual that at isolated points f is not smooth; it might even be discontinuous.

It is best to break up the problem into regions on which f is smooth, but it is common that users simply ignore the matter. There is quite a lot of evidence to the effect that a code based on EPUS cannot step past a severe lack of smoothness, but that a code based on EPS can. Further, although the cost of the integration and the accuracy achieved vary somewhat erratically, a code based on EPS is not greatly disturbed by isolated points where f is not smooth. See, for example, [2, 10, 11]. In this section we discuss some of the practical issues and apply the analysis of the preceding section to explain this experimental evidence for one kind of problem and one kind of method. The results indicate what might be expected in other situations.

The kind of problem we treat involves two expressions for f(x, y). We suppose that for $a \le x \le c$, we have a smooth function $f_1(x, y)$ and for $c \le x \le b$, we have another smooth function $f_2(x, y)$. By the solution of (1, 2) we mean a function y(x) defined in two pieces. For $a \le x \le c$, we have $y(x) = y_1(x)$ where

$$y'_1(x) = f_1(x, y_1(x)) \quad , \qquad y_1(0) = A \ .$$

For $c < x \le b$, we have $y(x) = y_2(x)$ where $y_2(x)$ is defined for $c \le x \le b$ by

$$y'_2(x) = f_2(x, y_2(x)) \quad , \qquad y_2(c) = y_1(c) \ .$$

The initial value problem is well-posed on [a, b] because small changes to A and f_1 lead to small changes in y_1 for $a \le x \le c$, in particular to small changes in $y_1(c)$. Further small changes to $y_2(c) = y_1(c)$ and small changes in f_2 lead to small changes in y_2 for $c < x \le b$. The lack of smoothness occurs along $x = c$. If it had occured along some other surface in the (x, y) space, the initial value problem on [a, b] might be ill-conditioned, or even fail to be well-defined. We have chosen the special problem for simplicity. It appears that the approach will succeed for other well-conditioned problems with isolated lack of smoothness in f.

The integration of $y_1(x)$ proceeds normally until a mesh point x_n is close enough to c that $c < x_n + h_n$. If the code can take a successful step to $x_n + h_n$, the integration of $y_2(x)$ will then proceed normally until the end point b is reached. This is not quite true of methods with memory because for them the lack of smoothness that troubled the step to x_{n+1} will affect several steps. One reason we restrict our attention to one-step methods is to avoid this complication.

Generally we suppose that f is smooth enough that the local error of our method of order p is in fact $0(h_n^{p+1})$. If f is not so smooth, this order is reduced to $0(h_n^{q+1})$ for some $q < p$. Unfortunately, the "higher order" formula used for estimating the local error will also have an error like $0(h_n^{q+1})$. Because both formulas are of the same effective order, the error estimate is not valid. The best we can say is that the difference we use to estimate the local error, est = $y^*_{n+1} - y_{n+1}$, is $0(h_n^{q+1})$, hence is of the same order as the local error. Of course, this is just a statement about the order of the error. It leaves open the very real possibility that for a particular step size, the estimate of the local error differs enormously from the local error itself.

It is illuminating to consider the extreme case of a jump discontinuity of f at $x = c$. Let $w(x)$ be defined by

$$w'(x) = f_1(x, w(x)) \qquad x \le c, \qquad w(x_n) = y_n$$

and

$$w'(x) = f_2(x, w(x)) \qquad x \ge c, \qquad w(c+) = w(c-) \ .$$

The local error of the step from x_n is then

$$w(x_n + h_n) - y_{n+1} \ .$$

Now it is clear that even when $x_n + h_n > c$,

$$w(x_n + h_n) = y_n + 0(h_n).$$

Also, for an explicit Runge-Kutta formula,

$$y_{n+1} = y_n + 0(h_n).$$

This implies that the local error is $0(h_n)$. The estimate of the local error, $y^*_{n+1} - y_{n+1}$, is also $0(h_n)$. Let us look at the estimate more closely. Now

$$y^*_{n+1} - y_{n+1} = h_n \sum \gamma_j f(x_{n,j}, y_{n,j}),$$

where the $x_{n,j}$ are points in $[x_n, x_n + h_n]$ and the $y_{n,j}$ approximate $y(x_{n,j})$. When h_n is small and f has a jump discontinuity at $x = c$, these $f(x_{n,j}, y_{n,j})$ are all approximately equal to either $f_1(c, y_1(c))$ or $f_2(c, y_1(c))$. Thus the multiple of h_n is approximately equal to one of a small set of possible values, which one depending on just where c is in the interval $[x_n, x_n + h_n]$ and the formula used. Clearly the value of the estimate is very sensitive to the details of the integration.

Notice that the estimated error does not go to zero faster than h_n when f has a discontinuity at $x = c$. This implies that it is generally not possible to step across such a point with EPUS because this criterion insists that

$$\| \text{local error estimate} \| \le h_n \tau \ .$$

In contrast, it is possible to pass such a point with EPS if the step size is taken sufficiently small.

What happens when a one-step code tries to step across $x = c$? It can happen that the lack of smoothness is not even noticed, but for small h_n (roughly equivalent to stringent tolerances), the step will fail. This is because the step size was predicted assuming that the problem will stay smooth. It is quite possible that the step fail very badly when there is a severe lack of smoothness at $x = c$. The usual step size adjustment algorithms need modification to deal with this situation. If the step does not fail too badly, the algorithm will suggest a modest reduction of the step size for the next attempt. The trouble is that it assumes that the order is p when in fact it is $q < p$, so that it does not reduce the step size enough. A standard device it to use a fixed fraction of the "optimal" step size. This helps in this situation, but repeated failures are common. If the step fails very badly, the algorithm will suggest a very much smaller step size for the next attempt. If this step does not reach $x = c$, we have taken a step which is very much smaller than is needed for the smooth problem. As a consequence, the step size algorithm suggests a very large increase which results in a spectacular failure as the code tries to step past $x = c$, and the cycle repeats. An example is given in [10]. An appropriate remedy is to limit the amount by which a step size can be decreased or increased. Limits on the change of step size are appropriate anyhow, because the asymptotic arguments that underlie step adjustment algorithms are not valid for large changes.

A device commonly seen is specifically intended to help with a severe lack of smoothness. Suppose that a step from x_n fails and after perhaps several tries a successful step size h_n is found. The device is to restrict the next step size to being no larger than h_n. The author has gone further in his most recent codes. Because step sizes are chosen conservatively, a step failure means that the problem is different that expected. We first reduce the step size in the usual way. This step size is also chosen conservatively, so that if it fails too, the problem is quite different from what was expected. At this point we begin halving the step size. Along with not increasing the step size after a failed attempt, this amounts to a bisection scheme for locating c and stepping over it with a step size about as large as possible.

We have seen that EPUS cannot deal with a jump discontinuity in f. When a code tries to cross such a discontinuity, it will keep reducing the step size until a test like that of [5] recognizes that h_n is too small to be meaningful in the precision of the computer, or too much work is done. The analysis of Section 2 says that a few steps taken with EPS will not destroy the attractive aspects of EPUS. If a code based on EPUS has several failed attempts to take a step from x_n, it should try using EPS for a couple of steps. This action should recognize and deal with a severe lack of smoothness.

At least for the kinds of discontinuous problems that we have investigated, our analysis says that we can expect a code based on EPS, or a suitably modified code based on EPUS, to solve the problems in a reasonable way. Although the error is not controlled as well as it is for a smooth problem, a *bound* has the same kind of behavior whether the problem is smooth or not. We emphasize "bound" here because the observed error will depend in a sensitive way on the details of the computation. It will, however, not differ greatly from the error that would be expected of a smooth problem. It is expensive to pass a point where f is not smooth, but only in a relative sense. We conclude, then, that the accuracy and cost of a substantial integration are not much affected by the presence of a few points where f is not smooth. This is why users have been able to ignore the presence of such points.

Hairer, Norsett and Wanner [4, p. 239] present some numerical results for a code DOPRI8 covered by our analysis. We shall take up their example below, and here wish only to comment about their results. They plot in Figure 10.4 the cost of the integration in evaluations of f against the global error at the end of the integration, and say " ... the obtained errors and number of function evaluations show a quite chaotic behavior". This is true of the fine structure, but the overall behavior is regular. The quantities plotted are not ideal for our purposes. As we have emphasized, both cost and accuracy are sensitive functions of the details of the integration, but they do not differ greatly from those expected of a smooth problem. This is why their plot shows points scattered in an erratic way about a smooth curve typical of the integration of a smooth problem.

5. LOCAL EXTRAPOLATION AND THE BDF.

It seems to be generally thought that local extrapolation of the BDF (backward differentiation formulas) is not practical. Byrne and Hindmarsh [1] explain why. They observe that when using, say, the BDF of order 3, adding in the estimated local error to get a result of order 4 yields a formula with poor stability. This is intolerable for the stiff problems that lead one to use a BDF code in the first place. The observation is correct, but what it means is that local extrapolation cannot be done in this way, not that it cannot be done at all.

A way around the difficulty is found when local extrapolation is viewed as arising from the use of two formulas. The error of the lower order one is used for step size control. The higher order formula is used to advance the integration. If, for example, we want the BDF3 as the lower order formula, we could choose the BDF4 as the higher. This choice provides a higher order formula of acceptable stability. There is an important practical matter that arises in the way these implicit formulas are evaluated. The conventional choice evaluates the implicit BDF3 and computes an approximation to the local error with some kind of Milne estimate, see, e.g. Gear [3]. The alternative we propose is to evaluate the implicit BDF4. It is easy to show that with this higher order result one can estimate the error of the BDF3 without having to evaluate this other implicit formula. The situation is essentially the same as that of Shampine and Gordon [9] for the Adams methods. It is not necessary even to introduce the BDF3 as a formula in its own right. One can just work with a difference approximation to its local truncation error computed from the higher order results.

Proceeding as we suggest it is not difficult to do local extrapolation with the BDF. Current practice at order 4 evaluates the implicit BDF4, estimates its error, selects a step size with EPS, and advances with the BDF4 result. Alternatively one can evaluate the implicit BDF4, estimate the error of the BDF3 and select a step size for it with EPS, and advance with the BDF4 result. The alternative is no more trouble and provides a more regular behavior with respect to a change of tolerance.

The real difference between EPS and XEPS in variable order BDF codes is seen at order 1. Conventional codes make use of the fact that the formula of order 1, the backward Euler method, is a one-step formula in order to start themselves. With XEPS as we have described it, the lowest order formula used is 2. This means that such codes have to be started in a different way. It is not difficult to accomplish this. However, the BDF1 is an effective formula, and it is not clear that dropping it is a good idea. In the next section we suggest a different way of getting the regular behavior of EPUS with respect to a change of tolerance without the disavantages of this way of implementing XEPS.

6. PSEUDO ERROR PER UNIT STEP.

For a given problem (1, 2) EPS is somewhat to be preferred to EPUS. In [7] two measures of efficiency are defined. As discussed more fully there, the two controls are equally efficient in one measure and EPS is more efficient in the other. In selecting the step size so that

$$h \beta \tau \approx \| le \| = h^{p+1} \| \phi \| + 0(h^{p+2})$$

for a method of order p, EPUS leads to

$$(9) \qquad h \approx (\beta \tau / \| \phi \|)^{1/p} .$$

Similarly EPS leads to

$$h \approx (\beta \, \tau / \| \phi \|)^{1/(p+1)} .$$

At a low order, EPUS leads to a significantly smaller step size. Popular Adams and BDF codes start at order 1. At this order there may be difficulties with EPUS that do not occur with EPS because EPUS needs a step size to small for the precision available. The same kind of thing happens when the effective order is reduced to one at a discontinuity of f and when τ is near limiting precision. The advantage in robustness enjoyed by EPS in these situations is not decisive, but has certainly contributed to the preference for EPS seen in popular codes.

Other issues arise when we think of solving a sequence of problems. It is common, indeed, that a problem be solved with a sequence of tolerances, the object being to assess the true accuracy by consistency. As we have seen in the bounds of Section 3 and by approximations in [7], when using EPUS the error is proportional to the tolerance τ. The behavior with EPS is a good deal less satisfactory. With a fixed, moderately high order, the behavior of EPS is not so different from EPUS. However, the behavior of a variable order code implementing orders as low as one is rather irregular. Some nice experiments by Enright [2] illustrate the behavior. On the other hand, EPUS is not invariant under a scale change of the independent variable, and EPS is (see [7]). This objection to EPUS is not nearly as important as the one above to EPS.

XEPS is a way to get the advantages of both EPS and EPUS. In the case of Runge-Kutta methods, the error estimate is sufficiently expensive that not using it for local extrapolation is a considerable waste. Largely for this reason local extrapolation is used in all the popular Runge-Kutta codes. The situation is different with Adams and BDF codes because the usual local error estimate is virtually "free". As we observed earlier, it has been thought that local extrapolation could not be done in a conventional BDF code. This is not true, but in view of the fact that we can estimate the local error at the current order when integrating at, say, order 4, it seems unnatural to control the step size according to the error at order 3. Here we present an alternative to XEPS that also secures the advantages of both EPS and EPUS. We call it pseudo error per unit step - PEPUS.

The idea we exploit is that the tolerance μ used inside the code need not be the one specified by the user, τ. The integration is to be done with EPS and a smaller tolerance μ. This yields all the advantages associated with the use of EPS for a given problem. To get the big advantage of EPUS on a change of tolerance, all we need to do is define

$$\mu = \tau^{(p+1)/p}$$

when at order p. Our bounds, and approximations in [7], say that with EPS at order p, the true error is proportional to

$$\mu^{p/(p+1)} = \tau .$$

In this way we get the proportionality with respect to the user's tolerance τ that we wanted. This is true even if the order is varied in the course of the integration. Because the steps are selected according to EPS, this new scheme is also invariant under a scale change of the independent variable.

The reason for the name PEPUS is that the step size is selected so that

$$\beta \mu \approx h^{p+1} \| \phi \| ,$$

hence

$$h \approx (\beta \mu / \| \phi \|)^{1/(p+1)} = \tau^{1/p} (\beta / \| \phi \|)^{1/(p+1)} .$$

The criterion of EPUS with tolerance τ leads to (9). Obviously the step sizes are selected in much the same way.

Although it appears that we have obtained all the advantages of EPS and EPUS, and none of the disadvantages, this is not quite true. The internal tolerance μ is more stringent than the given tolerance τ. How much more stringent depends on the order of the formula. At order 1, PEPUS calls for a step size just as small as EPUS does, as we observed, this can lead to precision difficulties. It should be appreciated that this is due to the order of the formula p, and not the effective order. There is no difficulty at a discontinuity where the effective order drops drastically because the step size is selected according to EPS. It should be further pointed out that the irregularity of the error as a function of the tolerance is not solely due to the error control. In the case of methods with memory, there are a variety of practical reasons for not adjusting the step size at every step like Runge-Kutta methods do. These reasons are especially important in BDF codes. This reluctance to change the step size means that the details of an integration do not depend on the tolerance in a smooth way. The most we can hope for is to improve the regularity of the error as a function of the tolerance on converting a BDF code from EPS to PEPUS. It is generally realized that conventional BDF codes deliver rather less accuracy than is requested via the local error tolerance. This is in large measure due to the use of EPS. It is cured by the more stringent internal tolerance implicit in PEPUS.

7. NUMERICAL EXAMPLES.

The problem (9.33) in Hairer, Norsett and Wanner [4] serves to illustrate numerically a number of the matters we have investigated in this paper. It is

(10) $$y' = -\operatorname{sign}(x)\,|1 - |x||\,y^2 , \qquad -2 \le x \le 2$$

(11) $$y(-2) = 2/3 .$$

The analytical solution of the scalar problem is easy enough to work out. It shows that y" has discontinuities at $x = \pm 1$ and y' has a discontinuity at $x = 0$. The solution does not vanish, so we used a pure relative error control of the local error in all our experiments. We performed our experiments with the RKF code of [8]. It is a Runge-Kutta code based on the Fehlberg (4,5) pair with error control by EPUS. Being a simple code, it is easily modified to experiment with different error criteria. A difficulty with any kind of tests involving costs is to account properly for the selection of the initial step size. We chose to give the code a value too large, and let it reduce the value to an appropriate one in its usual manner. We modified its counter of function evaluations so that we began measuring the work only after the code had found an appropriate initial value and initiated the integration. Proceeding in this way, all the variant codes had equally good starting values, and the observed costs reflect the difficulty of the problem rather than the quality of a guess for the starting step size. In all our experiments we examined the global error at each step and reported its maximum value. Our experience has been that the maximum value behaves more smoothly and is more useful for understanding a code's performance than the value at a single point. All the computations were performed in double precision on an IBM PC/AT.

By restricting the interval of integration to [-2, -1], we obtain a smooth problem. Table 1 presents the maximum global error and cost in function evaluations as a function of the input pure relative error tolerance for the three error criteria. At the cruder tolerances the codes take only one or two steps to integrate this easy problem. The behavior of the integration is then greatly disturbed by the adjustments of the step size necessary to produce a result at the end point. To reduce the distraction of this issue of output, we report in this table only results that cost at least 24 function evaluationas - four successful steps - and do similar things in other tables. We write, e.g., 4.1×10^{-5} as 4.1(-5). The code returns a warning message when it has done 3000 function evaluations. We terminated the run then. For this easy problem the behavior of the global error with respect to reducing the tolerance is as regular for EPUS as one could wish. It shows that the error itself is proportional to the tolerance τ, not just a bound for the error. PEPUS also performs extremely well in this respect. The behavior of EPS is typical, showing a proportionality to $\tau^{4/5}$. Notice that for a given tolerance τ, the code is trying to achieve (and does) a smaller error for EPUS and PEPUS than for EPS.

Integration of (10, 11) on [-2, 0] shows what happens when the order of the Runge-Kutta formula is reduced at x = -1 due to the jump in y" there. The results of Table 2 show that the behavior is a good deal less regular. All the variants of the code do a reasonable job of controlling the error and the cost is not greatly affected by the rough spot in the integration. Indeed, EPUS does a surprisingly good job of solving this problem.

Integrating (10, 11) on [-2, 2] is much more demanding. The truly bad spot is at x = 0 where y' is discontinuous, but there is an additional rough spot at x = +1 where y" is discontinuous. The fact that the severe lack of smoothness occurs at x = 0 is a bit awkward because x = 0 is a special point in the floating point number system. With EPUS the usual failure where f is discontinuous is due to h_n becoming too small for the precision available. At x = 0 the failure was always due to too much work. As Table 3 shows, this problem simply cannot be solved with EPUS. On the other hand, EPS and PEPUS do a reasonable job of controlling the error at a reasonable cost. The error reported for EPS at the tolerance 10^{-9} is not a mistake; the maximum error did get worse on reducing the imput tolerance by an order of magnitude. The regularity of the behavior of the error with PEPUS is surprisingly good. It is perhaps illuminating to realize that with this fixed order code, the integrations with EPS at the tolerances 10^{-5} and 10^{-10} are exactly the same as those with PEPUS at the tolerances 10^{-4} and 10^{-8}, respectively.

REFERENCES

[1] G.D.BYRNE and A.C.HINDMARSH. A polyalgorytm for the numerical solution of ordinary differential equations. ACM TOMS, 1 (1975) 71-96.

[2] W.H.ENRIGHT. Using a testing package for the automatic assessment of numerical methods for ODE's. pp. 199-217 in L.D.Fosdick, ed., Performance Evaluation of Numerical Software, North-Holland, Amsterdam, 1979.

[3] C.W.GEAR. Numerical Initial Value Problems in Ordinary Differential Equations. Prentice-Hall, Englewood Cliffs, NJ, 1971.

[4] E.HAIRER, S.P.NORSETT and G.WANNER. Solving Ordinary Differential Equations I Nonstiff Problems. Springer, Berlin, 1987.

[5] L.F.SHAMPINE. Limiting precision in differential equation solvers. Math. Comp., 27 (1974) 141-144.

[6] ______________. Local error control in codes for ordinary differential equations. Appl. Math. Comput., 3 (1977) 189-210.

[7] ______________. The step sizes used by one-step codes for ODEs. Appl. Numer. Math., 1 (1985) 95-106.

[8] ______________ and R.C.ALLEN, Jr., Numerical Computing: An Introduction. W.B.Saunders, Philadelphia, 1973.

[9] ______________ and M.K.GORDON. Computer Solution of Ordinary Differential Equations: the Initial Value Problem. W.H.Freeman, San Francisco, 1975.

[10] ______________ and H.A.WATTS, The art of writing a Runge-Kutta code. II, Appl. Math. Comput., 5 (1979) 93-121.

[11] ____________________________ and S.M.DAVENPORT, Solving nonstiff ordinary differential equations-the state of the art. SIAM Rev., 18 (1976) 376-411.

	EPUS		EPS		PEPUS	
Tolerance	error	cost	error	cost	error	cost
(- 5)	3.1(- 6)	29			3.5(- 7)	30
(- 6)	1.5(- 7)	47	6.3(- 7)	30	5.9(- 8)	65
(- 7)	2.0(- 8)	77	8.2(- 8)	53	7.1(- 9)	107
(- 8)	3.1(- 9)	131	2.7(- 8)	77	8.1(-10)	185
(- 9)	3.5(-10)	227	4.7(- 9)	119	8.5(-11)	323
(-10)	3.5(-11)	406	8.1(-10)	185	8.8(-12)	569
(-11)	3.6(-12)	707	1.4(-10)	287	9.0(-13)	1001
(-12)	3.6(-13)	1247	2.2(-11)	455	9.0(-14)	1775
(-13)	3.5(-14)	2207	3.5(-12)	713		*
(-14)		*	5.7(-13)	1121		*
(-15)		*	9.0(-14)	1775		*
(-16)		*	1.5(-14)	2807		*

Table 1. Solution of (10, 11) for $-2 \leq x \leq -1$. * means more than 3000 function evaluations.

Tolerance	EPUS error	EPUS cost	EPS error	EPS cost	PEPUS error	PEPUS cost
(-4)	3.2(- 3)	72			1.0(- 4)	74
(- 5)	9.8(- 5)	139	1.0(- 4)	74	2.1(- 5)	130
(- 6)	1.1(- 6)	214	1.3(- 4)	103	3.1(- 7)	207
(- 7)	3.0(- 7)	271	8.8(- 6)	203	1.8(- 7)	375
(- 8)	1.3(- 8)	540	2.4(- 7)	291	2.4(- 9)	593
(- 9)	1.9(- 9)	907	7.8(- 8)	394	8.8(-11)	908
(-10)	2.0(-10)	1244	2.4(- 9)	593	4.5(-11)	1525
(-11)	2.1(-11)	2127	2.8(-10)	775	4.8(-12)	2588
(-12)		*	1.2(-10)	1222		*
(-13)		*	2.0(-11)	1858		*
(-14)		*	2.4(-12)	2893		*

Table 2. Solution of (10, 11) for $-2 \le x \le 0$. * means more than 3000 function evaluations.

Tolerance	EPUS error	EPUS cost	EPS error	EPS cost	PEPUS error	PEPUS cost
(- 4)		*	2.2(- 2)	189	4.1(- 4)	327
(- 5)		*	4.1(- 4)	327	2.5(- 5)	431
(- 6)		*	5.7(- 5)	447	2.3(- 6)	705
(- 7)		*	3.1(- 6)	601	1.2(- 7)	972
(- 8)		*	2.1(- 7)	815	7.4(- 8)	1397
(- 9)		*	7.6(- 6)	1007	4.4(-10)	2195
(-10)		*	7.4(- 8)	1397	1.9(-11)	2996
(-11)		*	5.0(-10)	1927		*
(-12)		*	5.3(-11)	2826		*

Table 3. Solution of (10, 11) for $-2 \le x \le +2$. * means more than 3000 function evaluations.

LECTURE NOTES IN MATHEMATICS

Edited by A. Dold and B. Eckmann

Some general remarks on the publication of proceedings of congresses and symposia

Lecture Notes aim to report new developments - quickly, informally and at a high level. The following describes criteria and procedures which apply to proceedings volumes. The editors of a volume are strongly advised to inform contributors about these points at an early stage.

§1. One (or more) expert participant(s) of the meeting should act as the responsible editor(s) of the proceedings. They select the papers which are suitable (cf. §§ 2, 3) for inclusion in the proceedings, and have them individually refereed (as for a journal). It should not be assumed that the published proceedings must reflect conference events faithfully and in their entirety. Contributions to the meeting which are not included in the proceedings can be listed by title. The series editors will normally not interfere with the editing of a particular proceedings volume - except in fairly obvious cases, or on technical matters, such as described in §§ 2, 3. The names of the responsible editors appear on the title page of the volume.

§2. The proceedings should be reasonably homogeneous (concerned with a limited area). For instance, the proceedings of a congress on "Analysis" or "Mathematics in Wonderland" would normally not be sufficiently homogeneous.

One or two longer survey articles on recent developments in the field are often very useful additions to such proceedings - even if they do not correspond to actual lectures at the congress. An extensive introduction on the subject of the congress would be desirable.

§3. The contributions should be of a high mathematical standard and of current interest. Research articles should present new material and not duplicate other papers already published or due to be published. They should contain sufficient information and motivation and they should present proofs, or at least outlines of such, in sufficient detail to enable an expert to complete them. Thus resumes and mere announcements of papers appearing elsewhere cannot be included, although more detailed versions of a contribution may well be published in other places later.

Surveys, if included, should cover a sufficiently broad topic, and should in general not simply review the author's own recent research. In the case of surveys, exceptionally, proofs of results may not be necessary.

"Mathematical Reviews" and "Zentralblatt für Mathematik" require that papers in proceedings volumes carry an explicit statement that they are in final form and that no similar paper has been or is being submitted elsewhere, if these papers are to be considered for a review. Normally, papers that satisfy the criteria of the Lecture Notes in Mathematics series also satisfy this

.../...

requirement, but we would strongly recommend that the contributing authors be asked to give this guarantee explicitly at the beginning or end of their paper. There will occasionally be cases where this does not apply but where, for special reasons, the paper is still acceptable for LNM.

§4. Proceedings should appear soon after the meeeting. The publisher should, therefore, receive the complete manuscript within nine months of the date of the meeting at the latest.

§5. Plans or proposals for proceedings volumes should be sent to one of the editors of the series or to Springer-Verlag Heidelberg. They should give sufficient information on the conference or symposium, and on the proposed proceedings. In particular, they should contain a list of the expected contributions with their prospective length. Abstracts or early versions (drafts) of some of the contributions are very helpful.

§6. Lecture Notes are printed by photo-offset from camera-ready typed copy provided by the editors. For this purpose Springer-Verlag provides editors with technical instructions for the preparation of manuscripts and these should be distributed to all contributing authors. Springer-Verlag can also, on request, supply stationery on which the prescribed typing area is outlined. Some homogeneity in the presentation of the contributions is desirable.

Careful preparation of manuscripts will help keep production time short and ensure a satisfactory appearance of the finished book. The actual production of a Lecture Notes volume normally takes 6 -8 weeks.

Manuscripts should be at least 100 pages long. The final version should include a table of contents and as far as applicable a subject index.

§7. Editors receive a total of 50 free copies of their volume for distribution to the contributing authors, but no royalties. (Unfortunately, no reprints of individual contributions can be supplied.) They are entitled to purchase further copies of their book for their personal use at a discount of 33.3 %, other Springer mathematics books at a discount of 20 % directly from Springer-Verlag. Contributing authors may purchase the volume in which their article appears at a discount of 33.3 %.

Commitment to publish is made by letter of intent rather than by signing a formal contract. Springer-Verlag secures the copyright for each volume.

Vol. 1232: P.C. Schuur, Asymptotic Analysis of Soliton Problems. VIII, 180 pages. 1986.

Vol. 1233: Stability Problems for Stochastic Models. Proceedings, 1985. Edited by V.V. Kalashnikov, B. Penkov and V.M. Zolotarev. VI, 223 pages. 1986.

Vol. 1234: Combinatoire énumérative. Proceedings, 1985. Edité par G. Labelle et P. Leroux. XIV, 387 pages. 1986.

Vol. 1235: Séminaire de Théorie du Potentiel, Paris, No. 8. Directeurs: M. Brelot, G. Choquet et J. Deny. Rédacteurs: F. Hirsch et G. Mokobodzki. III, 209 pages. 1987.

Vol. 1236: Stochastic Partial Differential Equations and Applications. Proceedings, 1985. Edited by G. Da Prato and L. Tubaro. V, 257 pages. 1987.

Vol. 1237: Rational Approximation and its Applications in Mathematics and Physics. Proceedings, 1985. Edited by J. Gilewicz, M. Pindor and W. Siemaszko. XII, 350 pages. 1987.

Vol. 1238: M. Holz, K.-P. Podewski and K. Steffens, Injective Choice Functions. VI, 183 pages. 1987.

Vol. 1239: P. Vojta, Diophantine Approximations and Value Distribution Theory. X, 132 pages. 1987.

Vol. 1240: Number Theory, New York 1984–85. Seminar. Edited by D.V. Chudnovsky, G.V. Chudnovsky, H. Cohn and M.B. Nathanson. V, 324 pages. 1987.

Vol. 1241: L. Gårding, Singularities in Linear Wave Propagation. III, 125 pages. 1987.

Vol. 1242: Functional Analysis II, with Contributions by J. Hoffmann-Jørgensen et al. Edited by S. Kurepa, H. Kraljević and D. Butković. VII, 432 pages. 1987.

Vol. 1243: Non Commutative Harmonic Analysis and Lie Groups. Proceedings, 1985. Edited by J. Carmona, P. Delorme and M. Vergne. V, 309 pages. 1987.

Vol. 1244: W. Müller, Manifolds with Cusps of Rank One. XI, 158 pages. 1987.

Vol. 1245: S. Rallis, L-Functions and the Oscillator Representation. XVI, 239 pages. 1987.

Vol. 1246: Hodge Theory. Proceedings, 1985. Edited by E. Cattani, F. Guillén, A. Kaplan and F. Puerta. VII, 175 pages. 1987.

Vol. 1247: Séminaire de Probabilités XXI. Proceedings. Edité par J. Azéma, P.A. Meyer et M. Yor. IV, 579 pages. 1987.

Vol. 1248: Nonlinear Semigroups, Partial Differential Equations and Attractors. Proceedings, 1985. Edited by T.L. Gill and W.W. Zachary. IX, 185 pages. 1987.

Vol. 1249: I. van den Berg, Nonstandard Asymptotic Analysis. IX, 187 pages. 1987.

Vol. 1250: Stochastic Processes – Mathematics and Physics II. Proceedings 1985. Edited by S. Albeverio, Ph. Blanchard and L. Streit. VI, 359 pages. 1987.

Vol. 1251: Differential Geometric Methods in Mathematical Physics. Proceedings, 1985. Edited by P.L. García and A. Pérez-Rendón. VII, 300 pages. 1987.

Vol. 1252: T. Kaise, Représentations de Weil et GL_2 Algèbres de division et GL_n. VII, 203 pages. 1987.

Vol. 1253: J. Fischer, An Approach to the Selberg Trace Formula via the Selberg Zeta-Function. III, 184 pages. 1987.

Vol. 1254: S. Gelbart, I. Piatetski-Shapiro, S. Rallis. Explicit Constructions of Automorphic L-Functions. VI, 152 pages. 1987.

Vol. 1255: Differential Geometry and Differential Equations. Proceedings, 1985. Edited by C. Gu, M. Berger and R.L. Bryant. XII, 243 pages. 1987.

Vol. 1256: Pseudo-Differential Operators. Proceedings, 1986. Edited by H.O. Cordes, B. Gramsch and H. Widom. X, 479 pages. 1987.

Vol. 1257: X. Wang, On the C*-Algebras of Foliations in the Plane. V, 165 pages. 1987.

Vol. 1258: J. Weidmann, Spectral Theory of Ordinary Differential Operators. VI, 303 pages. 1987.

Vol. 1259: F. Cano Torres, Desingularization Strategies for Three-Dimensional Vector Fields. IX, 189 pages. 1987.

Vol. 1260: N.H. Pavel, Nonlinear Evolution Operators and Semigroups. VI, 285 pages. 1987.

Vol. 1261: H. Abels, Finite Presentability of S-Arithmetic Groups. Compact Presentability of Solvable Groups. VI, 178 pages. 1987.

Vol. 1262: E. Hlawka (Hrsg.), Zahlentheoretische Analysis II. Seminar, 1984–86. V, 158 Seiten. 1987.

Vol. 1263: V.L. Hansen (Ed.), Differential Geometry. Proceedings, 1985. XI, 288 pages. 1987.

Vol. 1264: Wu Wen-tsün, Rational Homotopy Type. VIII, 219 pages. 1987.

Vol. 1265: W. Van Assche, Asymptotics for Orthogonal Polynomials. VI, 201 pages. 1987.

Vol. 1266: F. Ghione, C. Peskine, E. Sernesi (Eds.), Space Curves. Proceedings, 1985. VI, 272 pages. 1987.

Vol. 1267: J. Lindenstrauss, V.D. Milman (Eds.), Geometrical Aspects of Functional Analysis. Seminar. VII, 212 pages. 1987.

Vol. 1268: S.G. Krantz (Ed.), Complex Analysis. Seminar, 1986. VII, 195 pages. 1987.

Vol. 1269: M. Shiota, Nash Manifolds. VI, 223 pages. 1987.

Vol. 1270: C. Carasso, P.-A. Raviart, D. Serre (Eds.), Nonlinear Hyperbolic Problems. Proceedings, 1986. XV, 341 pages. 1987.

Vol. 1271: A.M. Cohen, W.H. Hesselink, W.L.J. van der Kallen, J.R. Strooker (Eds.), Algebraic Groups Utrecht 1986. Proceedings. XII, 284 pages. 1987.

Vol. 1272: M.S. Livšic, L.L. Waksman, Commuting Nonselfadjoint Operators in Hilbert Space. III, 115 pages. 1987.

Vol. 1273: G.-M. Greuel, G. Trautmann (Eds.), Singularities, Representation of Algebras, and Vector Bundles. Proceedings, 1985. XIV, 383 pages. 1987.

Vol. 1274: N. C. Phillips, Equivariant K-Theory and Freeness of Group Actions on C*-Algebras. VIII, 371 pages. 1987.

Vol. 1275: C.A. Berenstein (Ed.), Complex Analysis I. Proceedings, 1985–86. XV, 331 pages. 1987.

Vol. 1276: C.A. Berenstein (Ed.), Complex Analysis II. Proceedings, 1985–86. IX, 320 pages. 1987.

Vol. 1277: C.A. Berenstein (Ed.), Complex Analysis III. Proceedings, 1985–86. X, 350 pages. 1987.

Vol. 1278: S.S. Koh (Ed.), Invariant Theory. Proceedings, 1985. V, 102 pages. 1987.

Vol. 1279: D. Ieşan, Saint-Venant's Problem. VIII, 162 Seiten. 1987.

Vol. 1280: E. Neher, Jordan Triple Systems by the Grid Approach. XII, 193 pages. 1987.

Vol. 1281: O.H. Kegel, F. Menegazzo, G. Zacher (Eds.), Group Theory. Proceedings, 1986. VII, 179 pages. 1987.

Vol. 1282: D.E. Handelman, Positive Polynomials, Convex Integral Polytopes, and a Random Walk Problem. XI, 136 pages. 1987.

Vol. 1283: S. Mardešić, J. Segal (Eds.), Geometric Topology and Shape Theory. Proceedings, 1986. V, 261 pages. 1987.

Vol. 1284: B.H. Matzat, Konstruktive Galoistheorie. X, 286 pages. 1987.

Vol. 1285: I.W. Knowles, Y. Saitō (Eds.), Differential Equations and Mathematical Physics. Proceedings, 1986. XVI, 499 pages. 1987.

Vol. 1286: H.R. Miller, D.C. Ravenel (Eds.), Algebraic Topology. Proceedings, 1986. VII, 341 pages. 1987.

Vol. 1287: E.B. Saff (Ed.), Approximation Theory, Tampa. Proceedings, 1985–1986. V, 228 pages. 1987.

Vol. 1288: Yu. L. Rodin, Generalized Analytic Functions on Riemann Surfaces. V, 128 pages, 1987.

Vol. 1289: Yu. I. Manin (Ed.), K-Theory, Arithmetic and Geometry. Seminar, 1984–1986. V, 399 pages. 1987.

Vol. 1290: G. Wüstholz (Ed.), Diophantine Approximation and Transcendence Theory. Seminar, 1985. V, 243 pages. 1987.

Vol. 1291: C. Mœglin, M.-F. Vignéras, J.-L. Waldspurger, Correspondances de Howe sur un Corps p-adique. VII, 163 pages. 1987

Vol. 1292: J.T. Baldwin (Ed.), Classification Theory. Proceedings, 1985. VI, 500 pages. 1987.

Vol. 1293: W. Ebeling, The Monodromy Groups of Isolated Singularities of Complete Intersections. XIV, 153 pages. 1987.

Vol. 1294: M. Queffélec, Substitution Dynamical Systems – Spectral Analysis. XIII, 240 pages. 1987.

Vol. 1295: P. Lelong, P. Dolbeault, H. Skoda (Réd.), Séminaire d'Analyse P. Lelong – P. Dolbeault – H. Skoda. Seminar, 1985/1986. VII, 283 pages. 1987.

Vol. 1296: M.-P. Malliavin (Ed.), Séminaire d'Algèbre Paul Dubreil et Marie-Paule Malliavin. Proceedings, 1986. IV, 324 pages. 1987.

Vol. 1297: Zhu Y.-l., Guo B.-y. (Eds.), Numerical Methods for Partial Differential Equations. Proceedings. XI, 244 pages. 1987.

Vol. 1298: J. Aguadé, R. Kane (Eds.), Algebraic Topology, Barcelona 1986. Proceedings. X, 255 pages. 1987.

Vol. 1299: S. Watanabe, Yu.V. Prokhorov (Eds.), Probability Theory and Mathematical Statistics. Proceedings, 1986. VIII, 589 pages. 1988.

Vol. 1300: G.B. Seligman, Constructions of Lie Algebras and their Modules. VI, 190 pages. 1988.

Vol. 1301: N. Schappacher, Periods of Hecke Characters. XV, 160 pages. 1988.

Vol. 1302: M. Cwikel, J. Peetre, Y. Sagher, H. Wallin (Eds.), Function Spaces and Applications. Proceedings, 1986. VI, 445 pages. 1988.

Vol. 1303: L. Accardi, W. von Waldenfels (Eds.), Quantum Probability and Applications III. Proceedings, 1987. VI, 373 pages. 1988.

Vol. 1304: F.Q. Gouvêa, Arithmetic of p-adic Modular Forms. VIII, 121 pages. 1988.

Vol. 1305: D.S. Lubinsky, E.B. Saff, Strong Asymptotics for Extremal Polynomials Associated with Weights on ℝ. VII, 153 pages. 1988.

Vol. 1306: S.S. Chern (Ed.), Partial Differential Equations. Proceedings, 1986. VI, 294 pages. 1988.

Vol. 1307: T. Murai, A Real Variable Method for the Cauchy Transform, and Analytic Capacity. VIII, 133 pages. 1988.

Vol. 1308: P. Imkeller, Two-Parameter Martingales and Their Quadratic Variation. IV, 177 pages. 1988.

Vol. 1309: B. Fiedler, Global Bifurcation of Periodic Solutions with Symmetry. VIII, 144 pages. 1988.

Vol. 1310: O.A. Laudal, G. Pfister, Local Moduli and Singularities. V, 117 pages. 1988.

Vol. 1311: A. Holme, R. Speiser (Eds.), Algebraic Geometry, Sundance 1986. Proceedings. VI, 320 pages. 1988.

Vol. 1312: N.A. Shirokov, Analytic Functions Smooth up to the Boundary. III, 213 pages. 1988.

Vol. 1313: F. Colonius, Optimal Periodic Control. VI, 177 pages. 1988.

Vol. 1314: A. Futaki, Kähler-Einstein Metrics and Integral Invariants. IV, 140 pages. 1988.

Vol. 1315: R.A. McCoy, I. Ntantu, Topological Properties of Spaces of Continuous Functions. IV, 124 pages. 1988.

Vol. 1316: H. Korezlioglu, A.S. Ustunel (Eds.), Stochastic Analysis and Related Topics. Proceedings, 1986. V, 371 pages. 1988.

Vol. 1317: J. Lindenstrauss, V.D. Milman (Eds.), Geometric Aspects of Functional Analysis. Seminar, 1986–87. VII, 289 pages. 1988.

Vol. 1318: Y. Felix (Ed.), Algebraic Topology – Rational Homotopy. Proceedings, 1986. VIII, 245 pages. 1988

Vol. 1319: M. Vuorinen, Conformal Geometry and Quasiregular Mappings. XIX, 209 pages. 1988.

Vol. 1320: H. Jürgensen, G. Lallement, H.J. Weinert (Eds.), Semigroups, Theory and Applications. Proceedings, 1986. X, 416 pages. 1988.

Vol. 1321: J. Azéma, P.A. Meyer, M. Yor (Eds.), Séminaire de Probabilités XXII. Proceedings. IV, 600 pages. 1988.

Vol. 1322: M. Métivier, S. Watanabe (Eds.), Stochastic Analysis. Proceedings, 1987. VII, 197 pages. 1988.

Vol. 1323: D.R. Anderson, H.J. Munkholm, Boundedly Controlled Topology. XII, 309 pages. 1988.

Vol. 1324: F. Cardoso, D.G. de Figueiredo, R. Iório, O. Lopes (Eds.), Partial Differential Equations. Proceedings, 1986. VIII, 433 pages. 1988.

Vol. 1325: A. Truman, I.M. Davies (Eds.), Stochastic Mechanics and Stochastic Processes. Proceedings, 1986. V, 220 pages. 1988.

Vol. 1326: P.S. Landweber (Ed.), Elliptic Curves and Modular Forms in Algebraic Topology. Proceedings, 1986. V, 224 pages. 1988.

Vol. 1327: W. Bruns, U. Vetter, Determinantal Rings. VII,236 pages. 1988.

Vol. 1328: J.L. Bueso, P. Jara, B. Torrecillas (Eds.), Ring Theory. Proceedings, 1986. IX, 331 pages. 1988.

Vol. 1329: M. Alfaro, J.S. Dehesa, F.J. Marcellan, J.L. Rubio de Francia, J. Vinuesa (Eds.): Orthogonal Polynomials and their Applications. Proceedings, 1986. XV, 334 pages. 1988.

Vol. 1330: A. Ambrosetti, F. Gori, R. Lucchetti (Eds.), Mathematical Economics. Montecatini Terme 1986. Seminar. VII, 137 pages. 1988.

Vol. 1331: R. Bamón, R. Labarca, J. Palis Jr. (Eds.), Dynamical Systems, Valparaiso 1986. Proceedings. VI, 250 pages. 1988.

Vol. 1332: E. Odell, H. Rosenthal (Eds.), Functional Analysis. Proceedings, 1986–87. V, 202 pages. 1988.

Vol. 1333: A.S. Kechris, D.A. Martin, J.R. Steel (Eds.), Cabal Seminar 81–85. Proceedings, 1981–85. V, 224 pages. 1988.

Vol. 1334: Yu.G. Borisovich, Yu. E. Gliklikh (Eds.), Global Analysis – Studies and Applications III. V, 331 pages. 1988.

Vol. 1335: F. Guillén, V. Navarro Aznar, P. Pascual-Gainza, F. Puerta, Hyperrésolutions cubiques et descente cohomologique. XII, 192 pages. 1988.

Vol. 1336: B. Helffer, Semi-Classical Analysis for the Schrödinger Operator and Applications. V, 107 pages. 1988.

Vol. 1337: E. Sernesi (Ed.), Theory of Moduli. Seminar, 1985. VIII, 232 pages. 1988.

Vol. 1338: A.B. Mingarelli, S.G. Halvorsen, Non-Oscillation Domains of Differential Equations with Two Parameters. XI, 109 pages. 1988.

Vol. 1339: T. Sunada (Ed.), Geometry and Analysis of Manifolds. Procedings, 1987. IX, 277 pages. 1988.

Vol. 1340: S. Hildebrandt, D.S. Kinderlehrer, M. Miranda (Eds.), Calculus of Variations and Partial Differential Equations. Proceedings, 1986. IX, 301 pages. 1988.

Vol. 1341: M. Dauge, Elliptic Boundary Value Problems on Corner Domains. VIII, 259 pages. 1988.

Vol. 1342: J.C. Alexander (Ed.), Dynamical Systems. Proceedings, 1986–87. VIII, 726 pages. 1988.

Vol. 1343: H. Ulrich, Fixed Point Theory of Parametrized Equivariant Maps. VII, 147 pages. 1988.

Vol. 1344: J. Král, J. Lukeš, J. Netuka, J. Veselý (Eds.), Potential Theory – Surveys and Problems. Proceedings, 1987. VIII, 271 pages. 1988.

Vol. 1345: X. Gomez-Mont, J. Seade, A. Verjovski (Eds.), Holomorphic Dynamics. Proceedings, 1986. VII, 321 pages. 1988.

Vol. 1346: O. Ya. Viro (Ed.), Topology and Geometry – Rohlin Seminar. XI, 581 pages. 1988.

Vol. 1347: C. Preston, Iterates of Piecewise Monotone Mappings on an Interval. V, 166 pages. 1988.

Vol. 1348: F. Borceux (Ed.), Categorical Algebra and its Applications. Proceedings, 1987. VIII, 375 pages. 1988.

Vol. 1349: E. Novak, Deterministic and Stochastic Error Bounds in Numerical Analysis. V, 113 pages. 1988.